T0134847

Rice By-products: Phytochemicals and Food Products Application

Bee Ling Tan • Mohd Esa Norhaizan

Rice By-products: Phytochemicals and Food Products Application

Bee Ling Tan
Department of Nutrition and Dietetics
Faculty of Medicine and Health Sciences
Universiti Putra Malaysia
Serdang, Selangor, Malaysia

Mohd Esa Norhaizan
Department of Nutrition and Dietetics
Faculty of Medicine and Health Sciences
Universiti Putra Malaysia
Serdang, Selangor, Malaysia

ISBN 978-3-030-46155-3 ISBN 978-3-030-46153-9 (eBook)
https://doi.org/10.1007/978-3-030-46153-9

This Springer imprint is published by the registered company Springer Nature Switzerland AG
The registered company address is: Gewerbestrasse 11, 6330 Cham, Switzerland

Preface

The prevalence of obesity has doubled from 1975 to 2016 worldwide. Approximately 1.5 billion people worldwide are overweight or obese, which increases the risk of developing type 2 diabetes, cancer, inflammatory disturbances, and cardiovascular disease. Despite numerous efforts and major advancements made to control these metabolic disorders, significant deficiencies and gap for improvements remain. The plant is of particular interest because plant-derived components modulate oxidative stress and hence alter gene and protein expression. Rice is one of the vitally important staple foods for almost half of the population in the world. The demand for rice is expected to remain strong over the next few decades due to the rapid growth of the population. Hence, the rice industry will remain sustainable for a long time, and the production of rice by-products including rice husk, rice straw, broken rice, rice germ, rice bran, and brewers' rice will remain high. Empirical evidence suggests that rice by-products may possess beneficial effects against oxidative stress and may favor for the prevention of metabolic disorders. These beneficial effects have been associated with the phytochemicals of the rice by-products present, such as vitamin E, dietary fiber, γ-oryzanol, γ-aminobutyric acid (GABA), and phytosterols. Based on our knowledge, the literature pertaining to rice by-products and its derived components with their molecular mechanisms that modulate non-communicable diseases has not well been compiled in the form of brief/book. Indeed, this disintegrated information needs to be compiled together to deliver knowledge at one point. Therefore, this book attempts to discuss issues pertaining to rice by-products, namely rice demands and rice by-products production, phytonutrients and antioxidant properties of rice by-products, potential health benefits, application in food products, and future prospects. By summarizing all the information in the lucid and comprehensive manner in one brief/book, it would provide a cohesive representation of the literature on the underlying mode of action involved in the pharmacological effect of these bioactive constituents that present in the rice by-products as well as plausible means for the prevention of metabolic ailments for the allied stakeholders and readers.

Serdang, Selangor, Malaysia

Bee Ling Tan

Abbreviations

5-LOX	5-Lipoxygenase
ABC-A	ATP-binding cassette
ACF	Aberrant crypt foci
AGEs	Advanced glycosylation end products
$AlCl_3$	Aluminum chloride
AOM	Azoxymethane
APC	Adenomatous polyposis coli
AR	Aldose reductase
ATP	Adenosine triphosphate
BHA	Butylated hydroxyanisole
BHT	Butylated hydroxytoluene
BMI	Body mass index
BOP	*N*-Nitrosobis(2-oxopropyl)amine
CE	Catechin equivalents
CEHC	γ-Carboxyethyl hydroxychroman
CK1	Casein kinase 1
COX	Cyclooxygenase
COX-2	Cyclooxygenase-2
CRP	C-Reactive protein
CVD	Cardiovascular disease
DBP	Diastolic blood pressure
DC-STAMP	Dendritic cell-specific transmembrane protein
DPPH	1,1-Diphenyl-2-picryl-hydrazyl
DPP IV	Dipeptidyl peptidase IV
ERK	Extracellular regulated protein kinases
FAO	Food and Agriculture Organization
FRAP	Ferric reducing antioxidant power
GABA	γ-Aminobutyric acid
GAD	Glutamate decarboxylase
GAE	Gallic acid equivalents
GSK3β	Glycogen synthase kinase 3β

HDL-C	High-density lipoprotein cholesterol
HEBR	Hydrolysates produced by limited enzymatic broken rice
HepG2	Hepatocellular carcinoma
HO-1	Heme oxygenase-1
HOMA-IR	Homeostasis model assessment-insulin resistance
HPA	Hypothalamic pituitary adrenal
HPLC	High-performance liquid chromatography
HT-29	Colon cancer cell lines
Hs	High-sensitivity
ICR	Imprinting control region
IFN-γ	Interferon-gamma
IgA	Immunoglobulin A
IL-1β	Interleukin-1beta
IL-6	Interleukin-6
iNOS	Inducible nitric oxide synthase
IP6	Inositol hexaphosphate
IRRI	International Rice Research Institute
JNK	c-Jun *N*-terminal kinase
LDL	Low-density lipoprotein
LDL-C	Low-density lipoprotein cholesterol
LPS	Lipopolysaccharides
LRP6	Low-density lipoprotein receptor-related protein 6
MAP	Mean arterial pressure
MAPK	Mitogen-activated protein kinase
MCF-7	Breast carcinoma
MDA	Malondialdehyde
NFATc1	Nuclear factor of activated T-cells, cytoplasmic 1
NF-κB	Nuclear factor-kappa B
NK	Natural killer
NMRI	Naval Medical Research Institute
NO	Nitric oxide
Nrf2	NF-E2-related factor 2
PC-3	Prostate cancer cell
PG	Prostaglandin
PGC1α	Peroxisome proliferator-activated receptor gamma coactivator 1-alpha
PGE_2	Prostaglandin E2
PGH_2	Prostaglandin H_2
PKC	Protein kinase C
PPARγ	Peroxisome proliferator-activated receptor-γ
PTH	Parathyroid hormone
PUFA	Polyunsaturated fatty acids
RANKL	Receptor activator of nuclear factor κB ligand
RBO	Rice bran oil
RNS	Reactive nitrogen species

ROS	Reactive oxygen species
SBP	Systolic blood pressure
SCFA	Short-chain fatty acids
SiO_2	Silica
SOD	Superoxide dismutase
SREBPs	Sterol regulatory element-binding proteins
TBHQ	tert-Butylhydroquinone
TC	Total cholesterol
TEAC	Trolox equivalent antioxidant capacity
TG	Triglycerides
TLR	Toll-like receptors
TMR	Total mixed rations
TNF-α	Tumor necrosis factor-alpha
USDA	United States Department of Agriculture
XOS	Xylooligosaccharides

Contents

Chapter 1
Introduction and Background

Abstract Rice is a staple food for nearly half of the global population. Irrigated rice contributes to nearly 55% of the global harvested area, which is about 75% of global rice production (410 million tonnes of rice per year), which is 100 times more productive compared to upland rice. The yields of rice production depend on water supply, herbicides, and fertilizers. They are several rice by-products produced during the rice milling process including rice bran, rice husk, brewers' rice, and rice straw. However, most of the rice by-products were used as animal feed and some of them were not efficiently utilized. Unused rice by-products are usually burnt in the field and thus leading to a serious environmental problem. Substantial evidence shows that rice by-products contain a high amount of bioactive compounds, for instance, pigmented compounds, γ-oryzanol, vitamin E, flavonoids, and phenolic acids that possess potential health benefits such as anticancer, antimutagenic, and antioxidative activity. Therefore, this book attempts to discuss issues related to rice by-products, including rice demands and rice by-products production, phytonutrients and antioxidant properties of rice by-products, potential health benefits, application in food products, and future prospects.

Keywords Irrigated rice · Rice by-products · Upland rice

Rice (*Oryza sativa* L.) supports a large number of populations over millennia compared to other crops as it was domesticated from 8,000 to 10,000 years ago (Greenland 1997). Currently, rice is one of the important cereal crops in Asia consumed by nearly half of the world's population as their daily staple food (Ali 2018). The rice is cultivated in more than 110 countries (Sharif et al. 2014). According to rice production quality data (2016) estimated by FAO, nearly 740 million tonnes of rice is harvested globally and about 670 million tonnes, which is 90% is consumed and produced in Asia, particularly eastern, southern, and south-eastern regions (Peanparkdee and Iwamoto 2019). Rice contributes up to 40% of caloric intake in tropical Asia but some countries can achieve more than 65% (Fairhurst and Dobermann 2002). Table 1.1 shows the rice production, area, and yield.

Paddy rice production has been sustained for a longer period of time and thus is considered as one of the world's most productive and sustainable farming systems (Horgan et al. 2016). Based on the annual basis, irrigated rice is 100 times more

B. L. Tan, M. E. Norhaizan, *Rice By-products: Phytochemicals and Food Products Application*, https://doi.org/10.1007/978-3-030-46153-9_1

Table 1.1 Rice production, area, and yield (Food and Agriculture Organization of the United Nations (FAOSTAT) 2019a)

	1990			2017		
Countries	Production (tonnes)	Area harvested (hg/ha)	Yield (ha)	Production (tonnes)	Area harvested (hg/ha)	Yield (ha)
Asia	477,693,369	132,426,521	36,072	692,590,948	145,539,189	47,588
India	111,517,408	42,686,608	26,125	168,500,000	43,789,000	38,480
China	191,614,680	33,518,971	57,166	214,430,049	31,035,082	69,093
Malaysia	1,884,984	680,647	27,694	2,901,894	689,268	42,101
Africa	12,697,110	6,034,413	21,041	36,560,295	14,959,657	24,439
Europe	4,570,432	1,063,346	42,982	4,051,459	642,982	63,010
World	518,568,653	146,960,085	35,286	769,657,791	167,249,103	46,019

productive compared to upland rice, about 5 times more productive than rainfed rice, and more than 12 times productive than deep-water rice (Fairhurst and Dobermann 2002). Irrigated rice contributes to nearly 55% of the global harvested area, accounting for about 410 million tonnes (75%) of global rice production annually (Dobermann and Fairhurst 2000). It is predominantly centered in regions of tropic and humid subtropical climate (Satapathy et al. 2015; Bayer et al. 2015). Intensified paddy rice system accounts for large population density as well as rich cultures that have been developed along with the major river systems in Asia (Hossain et al. 2016; Talhelm and Oishi 2018). Therefore, rice culture has been recognized as the cornerstone for the development of economic, cultural, and social in Asia (Spangenberg et al. 2018).

In the middle of the last century, the yields of rice production have increased steadily when several methods were developed to control the water supply. The farmer also used the varieties adapted to specific agro-ecological conditions (Vo et al. 2018). In general, rice was adapted to fit a broad spectrum of growing conditions, from deep-water swamps to the uplands and from the equatorial tropics to the high altitudes of Japan. Therefore, this could be explained why it is possible to collect up to 80,000 local varieties over the last 35 years, which stored at the International Rice Research Institute (IRRI) germplasm (Fairhurst and Dobermann 2002).

Several factors have changed on the demands of rice including per capita incomes, changes in the price of rice relative to substitute crops, and population growth (Dang et al. 2019). In the early 1960s, it was witnessed a steady increase in Asia's per capita rice production from 132,560 tonnes in the early 1960s to nearly 449,399 tonnes in 2013 (Food and Agriculture Organization of the United Nations (FAOSTAT) 2019b). These data indicate that the rising per capita production is more than doubled during this period, implied the high demands of rice in Asian countries. According to IRRI senior economist Dr. Samarendu Mohanty, rice consumption and their needs are expected to increase to an additional 116 million tonnes by 2035 to feed the global population (Subramanian 2013). Increasing rice consumption was driven primarily by population growth in Africa, Latin America, and Asia (Muthayya et al. 2014).

Before the introduction of herbicides, tall rice plants have a greater likelihood to have a competitive advantage compared to weeds. Furthermore, tall varieties were preferred because farmers often used rice straw for other purposes such as mulch, animal bedding, and fuel. Furthermore, tall varieties are easy to harvest (Fairhurst and Dobermann 2002). In the late 1950s, advancements in nitrogen fertilizer manufacturing technology led to a reduction of nitrogen cost. However, tall rice plants were less responsive to nitrogen fertilizer due to their susceptibility to lodging (Fairhurst and Dobermann 2002). Improvements in water control, herbicides, and crop protection as well as combined with the cost-effective nitrogen fertilizer, and thus leading the breeders to select plants with stiff and short straw that were less prone to lodging and produced high harvest index (Fairhurst and Dobermann 2002).

One of the most momentous histories of rice production was the crossing of the Indonesian variety Peta and Taiwanese variety Dee-geo-woogen to produce IR8, and thus began the Green Revolution. Therefore, they have been remarkably increased in grain yields. Local governments with the support from international lending agencies have made tremendous efforts and investments to enhance the water control in irrigated rice systems and thereby increase the area planted to rice and improve the cropping intensity (crops/ha/year) (Fairhurst and Dobermann 2002). Collectively, the combination of these systems has allowed the rice production to keep pace with the demand driven by the increase of growth in world population over the last few decades (Fairhurst and Dobermann 2002).

In addition to the nitrogen fertilizer, phosphorus fertilizer was also needed before a substantial response to nitrogen. In general, a country with a high deposit of natural gas and oil is used to manufacture nitrogen fertilizers. Manufactured phosphorus fertilizer is either produced or imported from local or oversea phosphorus rock, phosphoric acid, and sulphuric acid (Fairhurst and Dobermann 2002). In addition to phosphorus and nitrogen fertilizers, potassium fertilizer has been used sparingly in Asia's rice fields. However, potassium fertilizer could be served as an important production factor in the area, for instance, past soil mining and rice straw removal field (Fairhurst and Dobermann 2002). Emerging evidence highlights the importance of potassium fertilizer in the relation of pest resistance and plant health (Shi et al. 2018).

Several by-products are produced during the rice milling process including rice bran (combined with germ), rice husk, brewers' rice, and rice straw. In general, about 1,100 tonnes of straw can be obtained from the paddy field per year. Furthermore, rice processing also generates low-value by-product such as bran and husk, contributes for about 10 and 20% of the total weight of rice, respectively (Butsat and Siriamornpun 2010). In general, the rice milling by-products are utilized as an ingredient for animal feed (Sharif et al. 2014) as well as other purposes such as bedding material. Indeed, some of them were not efficiently utilized (Rafe and Sadeghian 2017). The disposal of this rice by-product has led to an adverse outcome (Daifullah et al. 2003). The previous finding has found that unused rice by-products are usually burnt in the fields and thus leading to the economic waste and environmental problems for instance loss of soil moisture, smog formation, and air pollution (Chaudhary et al. 2016). Intriguingly, substantial evidence highlights that rice by-products consist of numerous bioactive constituents, for instance, pigmented compounds, γ-oryzanol, vitamin E, flavonoids, and phenolic acids, which exerts potential health benefits (Peanparkdee et al. 2018; Kim et al. 2019; Liu et al. 2019). For example, rice bran contains several beneficial compounds such as phenolic compounds, tocotrienols, tocopherols, γ-oryzanol, and sterols (Peanparkdee and Iwamoto 2019). Notably, some of the phenolic acids found in rice bran, for example, diferulate, ferulic, and p-coumaric acids are not present in a significant amount in vegetables and fruits (Adom and Liu 2002). In addition to γ-oryzanol, a mixture of lipophilic phytosterols composed of sterols or triterpene alcohols with ferulic acid ester, shows cholesterol-lowering and antioxidant activity (Daou and Zhang 2014; Rungratanawanich et al. 2018). Several phenolic compounds, tocotrienols, and tocopherols in rice by-products have potentially beneficial effects, such as anticancer, antimutagenic, and antioxidative activity that play a vital role in promoting health (Hall et al. 2019). Research evidence has demonstrated that the bioactive compounds do not uniformly present in a cereal grain but are mainly concentrated in the bran and husk layers (Gao et al. 2018; Castanho et al. 2019). Therefore, dietary intake of whole grain in regular meals is strongly recommended to maintain desirable health benefits beyond basic nutrition as well as decrease the risk of numerous metabolic ailments (Gong et al. 2018).

According to the best of authors' knowledge, the literature pertaining to rice by-products and its derived components with their molecular mechanisms that modulate non-communicable diseases has not well been compiled in the form of brief/book. Indeed, this disintegrated knowledge needs to be combined together to deliver maximum information at one point. Therefore, this book attempts to discuss issues related to rice by-products, including rice demands and rice by-products production, phytonutrients and antioxidant properties of rice by-products, potential health benefits, application in food products, and future prospects.

References

Adom KK, Liu RH (2002) Antioxidant activity of grains. J Agric Food Chem 50:6182–6187

Ali N (2018) Co-occurrence of citrinin and ochratoxin A in rice in Asia and its implications for human health. J Sci Food Agric 98:2055–2059

Bayer C, Zschornack T, Pedroso GM et al (2015) A seven-year study on the effects of fall soil tillage on yield-scaled greenhouse gas emission from flood irrigated rice in a humid subtropical climate. Soil Tillage Res 145:118–125

Butsat S, Siriamornpun S (2010) Antioxidant capacities and phenolic compounds of the husk, bran and endosperm of Thai rice. Food Chem 119:606–613

Castanho A, Lageiro M, Calhelha RC et al (2019) Exploiting the bioactive properties of γ-oryzanol from bran of different exotic rice varieties. Food Funct 10:2382–2389

Chaudhary AK, Singh V, Tewari M (2016) Utilization of waste agriculture byproduct to enhance the economy of farmers. Ind Res J Ext Educ 12:89–92

Daifullah A, Girgis B, Gad H (2003) Utilization of agro-residues (rice husk) in small waste water treatment plans. Mater Lett 57:1723–1731

Dang KB, Windhorst W, Burkhard B et al (2019) A Bayesian Belief Network-based approach to link ecosystem functions with rice provisioning ecosystem services. Ecol Indic 100:30–44
Daou C, Zhang H (2014) Functional and physiological properties of total, soluble, and insoluble dietary fibers derived from defatted rice bran. J Food Sci Technol 51:3878–3885
Dobermann A, Fairhurst T (2000) Rice: nutrient management and nutrient disorders. PPI/PPIC and IRRI, Singapore, p 162
Fairhurst TH, Dobermann A (2002) Rice in the global food supply. Better Crops Int 16:3–6
Food and Agriculture Organization of the United Nations (FAOSTAT) (2019a) Food balance–food balance sheets. http://www.fao.org/faostat/en/#compare. Accessed 3 Jul 2019
Food and Agriculture Organization of the United Nations (FAOSTAT) (2019b) Production–crops. http://www.fao.org/faostat/en/#compare. Accessed 4 Jul 2019
Gao Y, Guo X, Liu Y et al (2018) A full utilization of rice husk to evaluate phytochemical bioactivities and prepare cellulose nanocrystals. Sci Rep 8:10482
Gong L, Cao W, Chi H et al (2018) Whole cereal grains and potential health effects: involvement of the gut microbiota. Food Res Int 103:84–102
Greenland DJ (1997) The sustainability of rice farming. CAB International, Wallingford, p 273
Hall KT, Kenneth JEB, Mukamal KJ et al (2019) COMT and alpha-tocopherol effects in cancer prevention: gene-supplement interactions in two randomized clinical trials. J Natl Cancer Inst 111(7):684–694
Horgan FG, Ramal AF, Bernal CC et al (2016) Applying ecological engineering for sustainable and resilient rice production systems. Procedia Food Sci 6:7–15
Hossain MS, Hossain A, Sarkar MAR et al (2016) Productivity and soil fertility of the rice-wheat system in the High Ganges River Floodplain of Bangladesh is influenced by the inclusion of legumes and manure. Agric Ecosyst Environ 218:40–52
Kim G-H, Ju J-Y, Chung K-S et al (2019) Rice hull extract (RHE) suppresses adiposity in high-fat diet-induced obese mice and inhibits differentiation of 3T3-L1 preadipocytes. Nutrients 11:1162
Liu YQ, Strappe P, Shang WT et al (2019) Functional peptides derived from rice bran proteins. Crit Rev Food Sci Nutr 59:349–356
Muthayya S, Sugimoto JD, Montgomery S et al (2014) An overview of global rice production, supply, trade, and consumption. Ann N Y Acad Sci 1324:7–14
Peanparkdee M, Iwamoto S (2019) Bioactive compounds from by-products of rice cultivation and rice processing: extraction and application in the food and pharmaceutical industries. Trends Food Sci Technol 86:109–117
Peanparkdee M, Yamauchi R, Iwamoto S (2018) Characterization of antioxidants extracted from Thai Riceberry bran using ultrasonic-assisted and conventional solvent extraction methods. Food Bioprocess Technol 11:713–722
Rafe A, Sadeghian A (2017) Stabilization of Tarom and Domesiah cultivars rice bran: physicochemical, functional and nutritional properties. J Cereal Sci 74:64–71
Rungratanawanich W, Abate G, Serafini MM et al (2018) Characterization of the antioxidant effects of γ-oryzanol: involvement of the Nrf2 pathway. Oxidative Med Cell Longev 2018:2987249. 11p
Satapathy SS, Swain DK, Pasupalak S et al (2015) Effect of elevated [CO2] and nutrient management on wet and dry season rice production in subtropical India. Crop J 3:468–480
Sharif MK, Butt MS, Anjum FM et al (2014) Rice bran: a novel functional ingredient. Crit Rev Food Sci Nutr 54:807–816
Shi X, Long Y, He F et al (2018) The fungal pathogen Magnaporthe oryzae suppresses innate immunity by modulating a host potassium channel. PLoS Pathog 14:e1006878
Spangenberg JH, Beaurepaire AL, Bergmeier E et al (2018) The LEGATO cross-disciplinary integrated ecosystem service research framework: an example of integrating research results from the analysis of global change impacts and the social, cultural and economic system dynamics of irrigated rice production. J Paddy Water Environ 16:287–319
Subramanian V (2013) Global rice trade faces uncertainty in 2013. In: Rice today, pp 38–39

Talhelm T, Oishi S (2018) How rice farming shaped culture in Southern China. In: Uskul AK, Oishi S (eds) Socioeconomic environment and human psychology. Oxford University Press, New York, pp 53–76

Vo TBT, Wassmann R, Tirol-Padre A et al (2018) Methane emission from rice cultivation in different agro-ecological zones of the Mekong river delta: seasonal patterns and emission factors for baseline water management. Soil Sci Plant Nutr 64:47–58

Chapter 2
Rice Demands: A Brief Description

Abstract More than 90% of rice is consumed and produced in the Asian Region, contributing to 80% of world consumption and production. Among all the factors, rising incomes and urbanization are the predominant contributors' consumption patterns. It can be inferred that rice by-products may increase substantially by 2050. In this regard, the annual cereal production is estimated to increase from 2.1 tonnes to nearly 3 billion tonnes, suggesting that an increase in consumption per capita, rise in population, and the use of cereals for biofuels may increase the rice demands. Extensive farming causes environmental issues, water scarcity, and soil salinity. Therefore, increased productivity and yield from new production and technologies are needed. Furthermore, it is hard to elevate rice production to meet these demands due to the water, labor, and land constraints as well as competition from the fast-growing non-farm sectors. In this chapter, we will describe the rice demands and rice production in Asia. Collectively, rice is considered as the life-blood in the Asia-Pacific Region that produces approximately 90% of the world's rice, in which the demands of rice are expected to grow faster than production in most of the countries.

Keywords Rice demands · Rice production · Rice consumption

Rice is a staple food for nearly half of the world population (Chaudhari et al. 2018), contributing to about 20% of global caloric intake. Approximately 90% of the rice is produced and consumed in the Asian Region including Japan, Vietnam, Bangladesh, Indonesia, India, and China, and contributing 80% of the world's consumption and production (Bandumula 2018). The growth in rice demands depends on several factors including (1) the change of prices relative to substitute crops; (2) growth population; and (3) level per capita income (Hossain 1997).

Based on the FAO report in world agriculture for 2030/2050, rice consumption per capita has stabilized after the end of 1980 with a slight decrease in East and South Asia. However, it was increased in other regions of the world, such as developed countries (Moraes et al. 2014). For example, per capita consumption of rice in China is more than 95 kg/year; while in Brazil, it is more than 41 kg/year (EMBRAPA—Embrapa Clima Temperado 2005; Wailes and Chavez 2012).

B. L. Tan, M. E. Norhaizan, *Rice By-products: Phytochemicals and Food Products Application*, https://doi.org/10.1007/978-3-030-46153-9_2

Table 2.1 The yields and area of husked rice

	1961/1963	2005/2007	2050
Production (million t)	230	644	827
Harvested area (million ha)	118	158	155
Yield (tonnes ha^{-1})	1.9	4.1	5.3

Source: Alexandratos and Bruinsma (2012)

The production of rice is expected to increase until 2050, with the same harvested area, owing to an elevation of productivity (Table 2.1). Likewise, a study reported by Lim et al. (2013) also showed a steady increase in rice demands in the next few decades, thus the rice industry will remain strong in the coming years. It can be inferred that the rice by-products will increase substantially by 2050. Hence, the existing concern to explore applications and alternatives for the by-products produced from the rice industry remains critical.

The data from FAO shows that to meet the projected demands, in which the world population increased by two billion or more than nine billion people by 2050, the global agricultural production must increase by 60% based on 2005–2007 levels. Among all the factors, rising incomes and urbanization are the predominant contributors' consumption patterns (Moraes et al. 2014). Several emerging economic countries in Latin America, North Africa, and Asia was rapidly changes in patterns and levels of food consumption, in which coarse grains, tubers, and roots were gradually replaced by increased intakes of dairy products and meat, vegetable oils, sugar, rice, and wheat (Food and Agriculture Organization of the United Nations (FAO) 2013).

Over the last six years, food prices are more volatile than the previous few decades (He et al. 2014). According to FAO (2012), annual cereal production is projected to increase from 2.1 billion tonnes to approximately 3 billion tonnes. This finding indicates that a significant increase in consumption per capita, a rise in population, and the use of cereals for biofuels are the main drivers of contribution (Alexandratos and Bruinsma 2012). Therefore, the average of global rice yields must increase from 4.1 tonnes/ha to 5.3 tonnes/ha. In fact, it would be challenging to achieve such increases in production due to the urbanization, competition from cash crops, climatic vulnerability, salinization of irrigated areas, soil degradation, and limited water and land resources (He et al. 2014). Wheat, maize, and rice are the vitally important cereals, accounting for 99% of total Chinese cereal production; despite other cereals for example sorghum, barley, and foxtail millet are also produced (He et al. 2014). In addition, maize, rice, and wheat are the most common cultured crops worldwide. Among these crops, rice is predominantly consumed by human in low-income countries (Pandey et al. 2010). Rice is cultured in many tiny farms mainly to meet the family needs. Although the marketable surplus is small, the price may fluctuate with typhoons, droughts, and floods. Hence, maintaining the stability of price and self-sufficiency in production is crucial in Asian countries. Table 2.2 shows the quantity of rice produced in several countries from 1995 to 2013.

Table 2.2 The production quantity of rice in several countries from 1995 to 2013

Country	Harvested quantity (Million tonnes)				
	1995	2000	2005	2010	2013
Africa	9.6	11.3	13.2	17.3	18.6
America	19.4	21.0	24.1	24.4	23.9
Asia	333.1	363.8	382.8	422.6	449.3
India	76.9	85.0	91.8	96.0	106.1
Indonesia	33.1	34.6	36.1	44.3	47.5
Bangladesh	17.6	25.0	26.5	33.3	34.3
Vietnam	16.6	21.6	23.9	26.6	29.3
Malaysia	1.4	1.4	1.5	1.6	1.7
Europe	1.8	2.1	2.2	2.8	2.6
World	364.8	399.1	422.7	467.5	495.4

Source: Food and Agriculture Organization of the United Nations (FAO) (2019)

2.1 Rice Production

The Green Revolution between 1940s and the late 1960s led to an elevation of agriculture in the developing countries, primarily via the transfer of technology and research initiatives (Food and Agriculture Organization 2006). The populations in low-income countries increased by 90% between 1966 and 2000, and thus increased the paddy rice by 130% (Muthayya et al. 2014). Nearly 84% of the rice production is attributed to the modern farming technologies that can produce semi-dwarf and early-maturing rice varieties, as well as responsive to nitrogen fertilizers (Maclean et al. 2002; Khush 2004).

Asian countries produce 89% of the world's consumption of rice, in which India and China alone accounting for 55% of the production (Milovanovic and Smutka 2017). However, rice is not equally consumed throughout the regions, with more urbanized nations, for example, Japan experiencing per capita consumption of 65 kg, which is 4 times lower than overpopulation countries such as Bangladesh (258 kg) (Milovanovic and Smutka 2017). Continent-wide consumption has increased from 95 kg per capita (1961) to 107 kg per capita (2013) (Milovanovic and Smutka 2017). The increased of area under rice by 35% since 1960 has improved the yield of 143%, which allowed production to overtake population growth. This progress is especially prominent in China where the yield is raised by 230%, while rice production in other countries has increased primarily via area expansion. Extensive farming causes environmental issues, water scarcity, and soil salinity. Hence, increased productivity and yield via new production and technologies are the alternative approach (FAO 2000). The Asian population is growing at 1.8% per year. Thus, the demand for rice is expected to increase by 70% over the next three decades (Hossain 1997). When technical progress is getting out of the stream, low-income countries will have problems to sustain the self-sufficiency in rice production, while the high- and middle-income countries may find it difficult to sustain farmers' interest in rice cultivation (Hossain 1997). The projection of the population

Table 2.3 The projection of population in rice consuming and producing countries in Asia, 1995–2025

Country	Population (million) 1995	Annual growth rate (% per year)		Projected population (million) in 2025	Percentage increase 1995–2025
		1995–2000	2020–2025		
China	1199	0.9	0.5	1471	23
India	934	1.7	1.0	1370	47
Indonesia	192	1.4	0.8	265	38
Bangladesh	121	1.8	1.1	182	50
Vietnam	74.1	2.0	1.2	117	58
Thailand	60.5	1.3	0.7	80.8	34
Myanmar	46.8	2.1	1.1	72.9	56
Japan	125	0.3	−0.3	124	−1
Philippines	69.2	2.2	1.2	115	66
Republic of Korea	44.8	0.8	0.3	52.9	18
Pakistan	130	2.7	1.6	243	87
Asia (excluding China)	2244	1.8	1.1	3389	51

Source: FAO (2000)

in rice consuming and producing countries in Asia from 1995 to 2025 is summarized in Table 2.3. Taken together, rice is the life-blood of the Asia-Pacific Region, in which 56% of humanity lives consuming and producing up to 90% of the world's rice. The rice demand is expected to grow faster than production in most countries (FAO 2000).

References

Alexandratos N, Bruinsma J (2012) World agriculture towards 2030/2050: the 2012 revision. ESA working paper no 12-03. FAO, Rome

Bandumula N (2018) Rice production in Asia: key to global food security. Proc Nat Acad Sci Ind Sec B Biol Sci 88:1323–1328

Chaudhari PR, Tamrakar N, Singh L et al (2018) Rice nutritional and medicinal properties: a review article. J Pharmacogn Phytochem 7:150–156

EMBRAPA—Embrapa Clima Temperado (2005) Cultivo do Arroz Irrigado no Brasil. http://sistemasdeproducao.cnptia.embrapa.br/FontesHTML/Arroz/ArrozIrrigadoBrasil/index.htm. Accessed 12 Nov 2013

FAO (2000) Bridging the rice yield gap in the Asia-Pacific region. Papademetriou, M.K., Dent, F.J. and Herath, E.M. Ed. Food and Agriculture Organization of the United Nations Regional Office for Asia and the Pacific. Bangkok. http://www.fao.org/3/x6905e/x6905e04.htm. Accessed 21 Oct 2019

FAO (2012) How to feed the world in 2050. http://www.fao.org/fileadmin/templates/wsfs/docs/expert_paper/How_to_Feed_the_World_in_2050.pdf. Accessed 5 Jul 2019

Food and Agriculture Organization (2006) Rice international commodity profile, 5 August 2014. FAO, Rome. http://www.fao.org/fileadmin/templates/est/COMM_MARKETS_MONITORING/Rice/Documents/Rice_Profile_Dec-06.pdf
Food and Agriculture Organization of the United Nations (FAO) (2013) FAO statistical yearbook 2013. World Food and Agriculture/FAO, Rome
Food and Agriculture Organization of the United Nations (FAO) (2019) Food balance—commodity balances—crops primary equivalent. http://www.fao.org/faostat/en/#compare. Accessed 22 Jan 2020
He Z, Xia X, Peng S et al (2014) Meeting demands for increased cereal production in China. J Cereal Sci 59:235–244
Hossain M (1997) Rice supply and demand in Asia: a socioeconomic and biophysical analysis. In: Teng PS, Kropff MJ, ten Berge HFM et al (eds) Applications of systems approaches at the farm and regional levels volume 1. Systems approaches for sustainable agricultural development, volume 5. Springer, Dordrecht, pp 263–279
Khush G (2004) Harnessing science and technology for sustainable rice based production systems. Presented at the FAO rice conference, Rome, Italy, 12–13 February 2004
Lim JS, Manan ZA, Hashim H et al (2013) Towards an integrated, resource-efficient rice mill complex. Resour Conserv Recycl 75:41–51
Maclean JL, Dawe DC, Hardy B et al (eds) (2002) Rice almanac. CABI Publishing, Wallingford, p 253
Milovanovic V, Smutka L (2017) Asian countries in the global rice market. Acta Univ Agric Silvic Mendelianae Brun 65:679–688
Moraes CAM, Fernandes LJ, Calheiro D et al (2014) Review of the rice production cycle: byproducts and the main applications focusing on rice husk combustion and ash recycling. Waste Manag Res 32:1034–1048
Muthayya S, Sugimoto JD, Montgomery S et al (2014) An overview of global rice production, supply, trade, and consumption. Ann N Y Acad Sci 1324:7–14
Pandey S, Byerlee D, Dawe D et al (eds) (2010) Rice in the global economy: strategic research and policy issues for food security. Los Baños, International Rice Research Institute
Wailes EJ, Chavez EC (2012) World rice outlook—international rice baseline with deterministic and stochastic projections, 2012–2021. http://ageconsearch.umn.edu/bitstream/123203/2/March%202012%20World%20Rice%20Outlook_AgEconSearch_05–01–12%20final.pdf. Accessed 9 Dec 2013

Chapter 3
Production of Rice By-products

Abstract Rice industry produces large amounts of waste. Rice processing involves some streams of milling steps to produce several materials including rice bran, husk, straw, germ, broken rice, and brewers' rice. In developing countries, these materials are considered as by-products and usually discarded as waste or used in animal feed. Indeed, the compositions of by-products are largely depended on the efficiency of the milling system and type of rice. The milling process is the crucial stage in rice production as it contributes to the cooking, sensory, and nutritional quality of rice. With the improvement of rice processing technology and the development of the social economy, consumers are chasing delicacy and taste. As rice processing becomes more and more refined, the extraction of phytochemicals in rice under various conditions is worth to discuss further. This chapter provides different physicochemical processing treatment and their impact on phytochemicals in rice. The production of by-products from the rice milling process is also highlighted in this chapter.

Keywords Rice bran · Rice germ · Rice straw · Rice husk · Broken rice · Brewers' rice

The growing population and consumption remains the contributor of rice production. In particular, almost all produced rice is used for human consumption domestically (Milovanovic and Smutka 2017). Rice is considered as a semi-aquatic annual grass plant, including nearly 22 species of the genus *Oryza*. Among these species, about 20 of them are wild types (Tateoka 1963). Importantly, the other two species are used for human consumption, namely *Oryza sativa* and *Oryza glaberrima*. Rice cultivars originated from Southeast Asia, namely *Oryza sativa* L., are differs from *Oryza glaberrima* Steud. species that are cultivated in West Africa. For example, the total area of cultivation in Thailand is 56.3 million Rai (22.3 million acres), comprising 90% of white rice cultivars, while pigmented rice contributes 0.1% or 62,000 Rai (24,506 acres) (The Rice Department MoAaC 2018). The multitude of varieties is classified into three categories, namely javanica (thick and broad grains), japonica (short roundish grains), and indica (slender, flat grains). The javanica is common to

B. L. Tan, M. E. Norhaizan, *Rice By-products: Phytochemicals and Food Products Application*, https://doi.org/10.1007/978-3-030-46153-9_3

tropical islands such as Indonesia, indica to tropical regions, and japonica to temperate regions. Among the subgroup, the most common varieties are vitreous kernels. Another type of rice is waxy or glutinous rice, in which the kernels are chalky and opaque and their cooking characteristics are different from common rice (Luh and Mickus 1991).

Despite white rice is regarded as a key staple food globally, several countries in Southeast Asia consume pigmented cultivars, for instance, brown, purple, black, and red rice (Arianti and Oktarina 2018; Pratiwi and Purwestri 2017; Samsudin and Abdullah 2014; Ahmad Hanis et al. 2012). Some of the rice cultivars exert pigments in their seed coat and pericarp which display color of the pigments on the surface (Huang and Lai 2016). The color intensities of pigmented rice seem to be linked to the indicators of its bioactive constituents (Chinprahast et al. 2016; Chen et al. 2016; Surarit et al. 2015). Substantial evidence has revealed that consumption of colored rice reduced oxidative stress *in vivo* concomitantly with increased antioxidant activity *in vivo* and *in vitro* (Ling et al. 2001; Toyokuni et al. 2002; Xia et al. 2003). This finding highlights the roles of unique complexes of bioactive constituents in rice, such as polyphenols, tocols, and γ-oryzanols. The color of black and purple kernels is related to the presence of anthocyanins (Reddy et al. 1995).

3.1 Rice Processing

Rice industry is a vitally important field that produces large amounts of waste (Moraes et al. 2014). Rice grain needs to undergo several rice processing steps before it can be consumed by humans. The structure of rice grain is shown in Fig. 3.1. There are several rice processing steps including harvesting, transport, pre-cleaning, drying, storage, shelling, milling, and classification (Moraes et al. 2014).

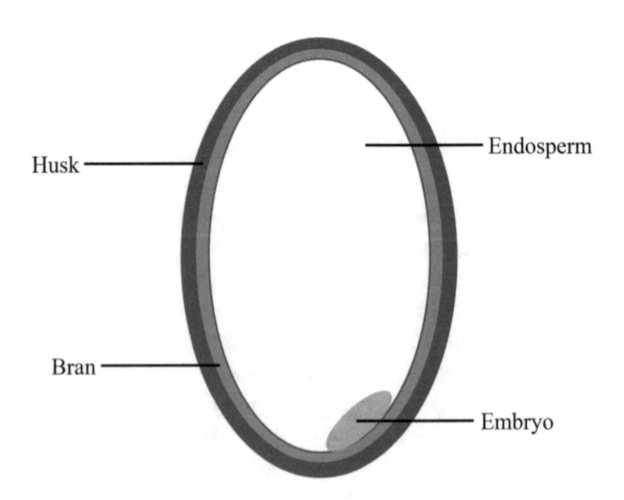

Fig. 3.1 Structure of rice grain

The rice processing and production covers from the operation of the harvest to the graded and polished white rice (Lim et al. 2013; Saidelles et al. 2012).

Rice milling process can be divided into three milling systems, namely multi-stage, two-step, and one-step milling (Odior and Oyawale 2011). The multistage milling system is usually applied at the commercial levels. This milling system involves an intricate system to reduce grain breakage (International Rice Research Institute (IRRI) 2016). The multistage rice milling process is summarized in Table 3.1.

During the rice milling process, the paddy or rough rice is initially cleaned to separate contaminants and particulates, and the husk is removed by shellers. The

Table 3.1 Rice milling process

Stage	Rice by-products	Description	References
Pre-cleaning	Rice straw	Foreign materials such as soil, weed seeds, straw, and stones are removed. Rice quality is depends on milling recovery	Dhankhar (2014); International Rice Research Institute (IRRI) (2016)
Dehusking/ dehulling	Rice husk	The removal process of the husk from rice paddy by friction	Dhankhar (2014); International Rice Research Institute (IRRI) (2016)
Paddy separation	Small broken rice, bran, and husk	The huller is used to remove the lighter materials, for example, small broken rice, bran, and husk. The remainder passes via the paddy separator to further separate broken rice from unhulled	International Rice Research Institute (IRRI) (2016)
Whitening/ polishing	Bran	The bran layer is removed from the germ and produces white rice. The amount of bran removed is usually 8–10% of the total paddy weight and varied according to the whiteness	International Rice Research Institute (IRRI) (2016)
Separation and grading of white rice	Broken rice and brewers' rice	After polishing, white rice is classified according to its size. Head rice comprises of 75–80% of the whole kernel. Rice that does not meet the size requirement is categorized as brewers' rice	Dhankhar (2014); International Rice Research Institute (IRRI) (2016)
Rice mixing	–	The head rice is blended with the correct portion of broken rice	International Rice Research Institute (IRRI) (2016)
Mist polishing	–	Polishing is conducted by a whitening machine using a mist of water to refine rice before it is ready for sale	Dhankhar (2014)
Rice weighing	–	Rice is sold in 50 kg sacks; however, the size of the bags depends on the requirements of the customer	Dhankhar (2014); International Rice Research Institute (IRRI) (2016)

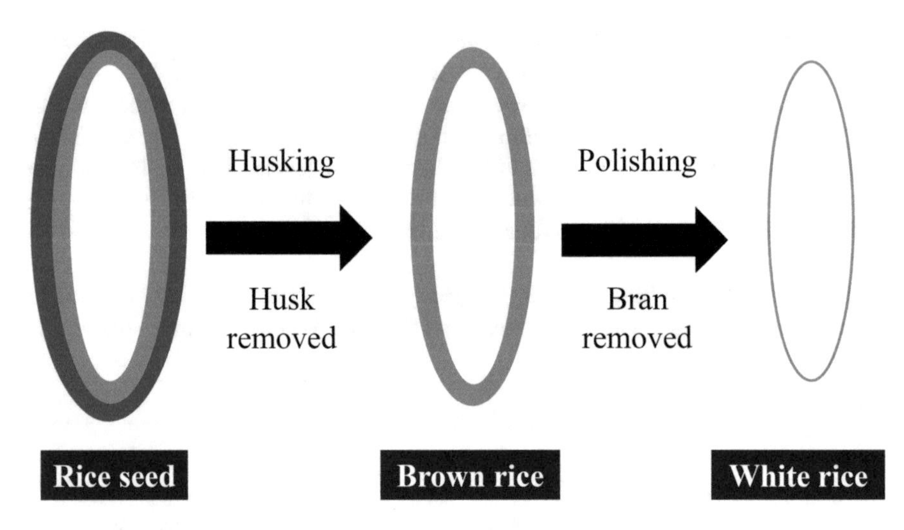

Fig. 3.2 Commercial rice processing

sheller is horizontally spaced rotating abrasive stones, in which rubber belt shellers or rubber-roll has been increasingly used in recent decades (Ajala and Gana 2015). The husks and brown rice are separated by aspiration and paddy remaining with the brown rice is separated using paddy separator (Prabhakaran et al. 2017). The brown rice goes to a milling machine, known as huller, in which the bran is abraded away to leave nearly white kernel (Liu et al. 2018). The rice is then passed through the brushes to remove a final layer of material known as polish (Fig. 3.2). The milled rice consists of broken and whole grains, namely whole kernels (head rice), small broken (brewers' rice and screenings), and large broken grains or broken (second heads) (Fig. 3.3). Rice grains are comprised of 69% starchy endosperm (milled rice), 11% rice bran, and about 20% rice husk (Dhankhar 2014). The endosperm fractions are typically packaged and sold for either further processing or direct consumption, while the rest of the rice kernel (31%) is rice by-products. In the past, rice by-products were discarded as waste (Sharif et al. 2014). Nonetheless, due to the nutrient composition and bioactive compounds, rice by-products may have industrial applications for animals and humans (Rohman et al. 2014).

3.2 Processing Treatments Affect the Phytochemical Contents of Rice

Phytochemicals are bioactive compounds found in whole grains, vegetables, fruits, and other plant foods. It can be categorized as organosulfur compounds, nitrogen-containing compounds, phenolics, and carotenoids (Chen and Liu 2018). Whole grains are rich in phytochemicals such as vitamin E, phenolics (syringic, vanillic, ferulic, caffeic, and p-coumaric acids), and carotenoids (β-carotene, β-cryptoxanthin,

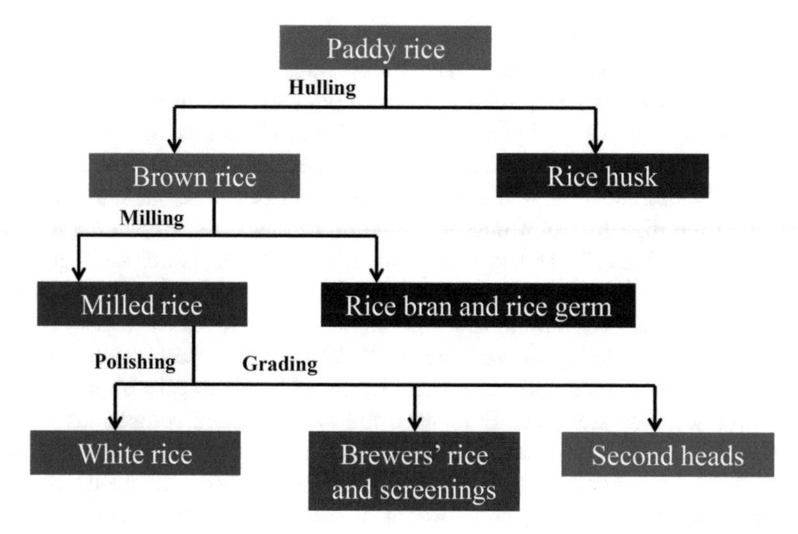

Fig. 3.3 Rice processing and resulting products

zeaxanthin, and lutein) (Zhu and Sang 2017). Phenolic acids are primarily found in the cortical layer of grains, in which the highest component is ferulic acid, followed by oxalic, p-coumaric, and caffeic acids (Juliano and Tuaño 2019). Phytochemicals are well-recognized for their wide range of health-promoting effects on animals and humans (Panche et al. 2016). The low redox potential of flavonoids decreased highly oxidized free radicals including peroxyl, hydroxyl, alkoxyl, and superoxide radicals by proton donation (Kovacic and Somanathan 2011). Phytochemicals or phytochemical-enriched by-products have been widely used in bakery foods due to their nutraceutical properties (Irakli et al. 2017). Although their potentials in the cosmetic and pharmaceutical industries are reported, many environmental factors such as oxygen, temperature, and light may influence the physicochemical stability of phytochemicals.

3.2.1 Physical Treatment

3.2.1.1 Effect of Grinding on Phytochemicals in Rice

Milling is a primary step in grain physical processing. It can be classified into two classes, namely wet grinding and dry grinding (Papageorgiou and Skendi 2018). Dry grinding separates germs and outer fibers; whereas wet grinding produces cereal endosperm products. Dry grinding comprised of two steps, which are screening and grinding. Screening predominantly refers to removing foreign particles, weeds, bacterial discolored grain, pest-infested grain, and impurities (Andrés et al. 2018). It has been shown that overmilling destroys the nutrients in grain (Tylewicz et al. 2018). Wet grinding has been utilized to separate fiber, protein, and starch by soaking the grains in water. However, poor formation capacity of gluten and high content of pentosan make it inapplicable for the industrial wet milling process (Heo et al. 2013). In addition, the protein cortex and bran layer are hard to separate completely, and high amounts of glucan may also lead to increased viscosity, and thus prevent it to pass through screening and centrifugation (Serna-Saldivar 2016).

During the milling process, the flavor and taste of rice can be markedly improved along with the digestibility. Conversely, the nutritional values of rice are greatly reduced (Liu et al. 2017). This is because the germ and bran, which are high in biologically bioactive compounds, minerals, and vitamins, are removed by milling (Smuda et al. 2018). The antioxidant compounds of rice are predominantly found in the germ and aleurone layers, and thus the milling process may lead to inevitably adverse outcomes to their antioxidant activity and phytochemicals. Improvement of processing accuracy significantly reduced the total phenolic compounds for both indica and japonica brown rice, where the combined phenol component was reduced to a greater extent compared to free phenol (Gong et al. 2017). Basically, the combined and free phenolic acid composition of brown rice is the same in different milling degrees; conversely, the phenolic acid is markedly different (Ding et al. 2019). Furthermore, the antioxidant activity of brown rice is also significantly reduced during the milling process. After milling for 30 s, the cellular antioxidant activity and total phenol content of brown rice reduced by 92.85 and 55.50%, respectively (Liu et al. 2015a). This is primarily due to the thermal effect caused by grinding and the removal of the skin high in phytochemicals and thus resulting in the oxidation of polyphenols (Liu et al. 2015b).

3.2.1.2 Effect of Thermal Processing on Phytochemicals in Rice

In the food industry, thermal processing may greatly affect the composition of proteins and bioactive compounds due to the several processing factors including copigments, oxygen, enzymes, thermal treatment, light, and pH (Tiwari et al. 2009). The study has shown that thermal treatment often leads to the denaturation of the native structure of food proteins, which is crucial for protein functionality. In particular, rice protein has very interesting properties, for example, hypoallergenic, good nutritional values, and healthful properties for human consumption (Baccaro et al. 2018). Despite rice proteins are healthful and hypoallergenic as well as contained high nutritional value, limited studies concerning their conformational and structural properties (Shin et al. 2011; Han et al. 2015; Thakur et al. 2015; Romani et al. 2017). In this regard, there are limited studies reported on the thermal denaturation of rice protein fractions as a food ingredient, for instance, baby formulas, ice creams, puddings, and gels (Bolea et al. 2019). In most Asian countries, there are several heat treatment methods including tempering, drying, baking, steaming, and boiling (Bolea et al. 2019). The effect of different heat treatment on phytochemicals in rice was evaluated (Pradeep and Guha 2011; Cardoso et al. 2014). Bolea et al. (2019) evaluated the kinetics of phytochemical thermal degradation in different milled fractions of black rice. The data revealed that anthocyanins being the most thermolabile compound that degrades at the highest rate, which significantly affects the antioxidant activity (Bolea et al. 2019). The previous data revealed that total antioxidant capacity in whole-kernel red rice was greatly reduced during water cooking, suggesting the loss of soluble and free antioxidants during cooking (Finocchiaro et al. 2007). Notably, water cooking increased the tocols levels of all the dehulled white and red rice (Finocchiaro et al. 2007). Heat treatments also release the bound tocols (tocol molecules that are strongly bound to the cellular compounds) (Qureshi et al. 2000).

Srichamnong et al. (2016) studied the effects of various cooking methods (frying, steaming, and boiling) on parboiled germinated brown rice, white rice, and brown rice on tocols and γ-oryzanol. Among all the cooking methods, steaming can retain the loss of γ-oryzanol (nearly 0% loss). The favorable outcome could be attributed to the gelatinization, which increased the starch viscosity that may reduce the loss of tocols and γ-oryzanol (Srichamnong et al. 2016). It has been demonstrated that γ-oryzanol is reduced during the boiling process, suggesting that this bioactive compound is leaching out into the water during boiling (Srichamnong et al. 2016). The data further demonstrated that boiled parboiled germinated brown rice has a higher amount of γ-oryzanol compared to boiled brown rice. Such findings implied the bran layer may act as a protective coat to γ-oryzanol and thereby reducing the leaching rate of γ-oryzanol in parboiled germinated brown rice compared to brown rice during boiling. Indeed, water plays a crucial role in rice cooking particularly during gelatinization to dissolution and swelling, and boiling with excessive water could increase the dissolution of tocols and γ-oryzanol, and thus reducing the amount of γ-oryzanol. In addition, the leaching of amylose and amylopectin also resulted in different degrees of gelatinization during rice boiling. Frying is considered as a harsh cooking procedure for γ-oryzanol in brown rice and parboiled germinated brown rice as it was decreased to half. The observed effect could be attributed to a high temperature (100 °C) during frying. In this regard, the oil was utilized as a heat transfer medium. The heat transfer rates for frying, boiling, and steaming were about 7.5, 4.5, and 6.67 °C/min, respectively. A high heat transfer rate promotes the degradation of γ-oryzanol (Srichamnong et al. 2016). In line with the effect observed in γ-oryzanol, steaming showed higher amounts of total tocols because gelatinization can protect tocols from the heat by remaining inside the grain (Srichamnong et al. 2016). Interestingly, boiling did not significantly degrade the amount of tocol as observed in γ-oryzanol, implied that tocols have different chemical structures and heat stability (Srichamnong et al. 2016). Compared to boiling and frying, steaming is one of the best cooking methods for preserving crucial bioactive compounds.

3.2.1.3 Effect of Extrusion on Phytochemicals in Rice

The extrusion of whole grain has become popular in the pharmaceutical and food industries (dos Santos et al. 2019; Anunciação et al. 2019). Extrusion is a versatile process comprises of several operations including forming, shaping, shearing, kneading, cooking, and mixing (Singh Gujral and Singh 2002; Xu et al. 2016). It changes the macroscopic shape, chemical characteristics, and microstructure of products (Hagenimana et al. 2006; Arribas et al. 2017). Extrusion can be divided into two groups, namely cold and hot extrusion, based on the type of construction (twin- or single-screw extruder) and the methods of operation (Mishra et al. 2012). During the mechanical process, the extruded materials exposed to high pressures, shear forces, and temperatures over a short period of time. High temperatures of the food material in the barrel of the extruder lead to the formation of complexes between proteins, lipids, and starch, denaturation of protein, and gelatinization of starch (Wolfe and Liu 2003). Extrusion is vitally important in the manufacturing process of food products such as cereal-based baby food, precooked flours, crispy flatbread, savory snacks, and breakfast cereals (Torbica et al. 2019). It affects the nutritional components and quality of extruded products. The phenolic compounds may also undergo several changes and thus changing the antioxidant activity in the food (Garzón et al. 2019). Carbohydrates undergo major changes among all the macronutrients. Starch is one of the predominant carbohydrates and crucial structural components in a variety of rice products (Witono and Juliani 2019). In particular, starch is given a great deal of interest because it enhances the quality of the products as well as the process convenience (Zheng et al. 2018). The primary changes in starch caused by extrusion cooking including depolymerization, dextrinization, and gelatinization (Neder-Suárez et al. 2018). The extrusion process can further destroy the structure and crystallinity of starch molecules (Chinnaswamy and Hanna 1990). The gelatinization of starch increased gas-holding capacity and expansion of extruded products (Horstmann et al. 2017). The combination of screw speed, moisture content, and temperature decreases the amounts of carbohydrate from 53% in the native flour to 1–8% in the final extruded materials due to the breakdown of amylopectin into intermediate molecular weight products (Chanvrier et al. 2015). The extrusion process altered the starch qualitatively and the addition of materials rich in fibers and proteins further improved the nutritional quality of the rice-based extruded products (Nascimento et al. 2017). The molten starch sticks to the cellulosic wall during extrusion of low amylose rice flour and thus leading to the formation of a complex wall that hindered the expansion (Borah et al. 2016). Increasing total starch levels in the rice extrudate could be attributed to the formation of reducing sugars due to the severe shearing (Arribas et al. 2017). Rathod and Annapure (2017) found that lentil-based rice noodles have higher digestibility of starch compared to the protein during extrusion, suggesting that legume starch structure may play a crucial role in the digestion of starch.

Phytochemicals in processed grain and grain samples are distributed in rice inbound, soluble-conjugated, and free forms. The majority of the bound and free compounds are present in different parts of grains, especially in the distinct portions obtained from milling the grains (Onyeneho and Hettiarachehy 1992). The extrusion treatment facilitates the extraction of free phenol, enhances the transformation of binding polyphenols to free phenol, and destroys the cell wall (Tian et al. 2019). Heat compression caused a partial degradation of polyphenols as well as changes in molecular structure and led to a certain degree of polymerization (Tian et al. 2019).

It is worth noting that when the water increased more than 18%, the polymerization of polyphenols is increased, and thus resulted in a reduction of antioxidant activity (Tian et al. 2019). However, the low water content in the feed (less than 15%) accelerated the depolymerization of condensed tannin and converted into low molecular weight oligomers that can be easily extracted. Moreover, extrusion also increases or decreases the total phenolic content and its monomer phenol in the grain and thus affects the antioxidant activity (Tian et al. 2019).

A previous study showed that extrusion of rice leads to the retention of several antioxidant components (Ohtsubo et al. 2005). Extrusion of brown rice flours (Sharbati, PR-106, and IR-8) at 100 °C significantly decreased the antioxidant activity and total phenolic content (Gujral et al. 2012). This finding is in line with the earlier study who found that the extrusion process with the screw speed (300, 350, and 400 rpm) and feed moisture (14, 18, and 22%) significantly decreased the antioxidant activity of extruded snacks from germinated brown rice flour (Chalermchaiwat et al. 2015). In addition, the total phenolic content is also affected by the screw speed and feed moisture content (Lohani and Muthukumarappan 2017). High screw speed and low moisture content significantly retained the total phenolic content (Chalermchaiwat et al. 2015). A study by Yağcı and Göğüş (2008) found that extrusion of rice-based snacks at temperature 150–175 °C can significantly increase the total phenolics at 12–18% moisture content. Increased moisture content at low temperatures during enzymatic extrusion of rice flour further enhanced the total phenolic content (Xu et al. 2016). The observed effect could be attributed to the increased enzymatic activity at higher moisture that created a mild extrusion environment. Studies have shown that extrusion increases the free chlorogenic acid (27.1%) and decreases the free gallic acid (45%) (Ti et al. 2015). From the study reviewed, extrusion leads to a significant loss of bound phenolic acid (Ti et al. 2015). The total and soluble free phenolic acids are markedly increased with the transformation of the soluble conjugated and soluble free phenolic acids into insoluble bound phenolic acid when increased in temperature and decreased in pressure (Hu et al. 2018).

3.2.2 *Chemical Treatment*

3.2.2.1 Effect of Germination on Phytochemicals in Rice

Germination is a complex physiological process, in which the endogenous enzymes are stimulated and released, and thus leading to the recombination and decomposition of the internal components of grains. This process may cause a certain impact on their phytochemicals content and antioxidant activity (Paucar-Menacho et al. 2017). During the grain germination process, the sensory, nutritional, chemical, and biological properties of rice may change significantly. For example, the soluble protein was increased and certain enzymes in the rice are activated. Several studies reported by Omary et al. (2012) and Peanparkdee et al. (2019) have found that the germination process can promote the polyphenols content in grains such as brown rice and millet.

Germination can affect the nutritional content in grains such as vitamins, minerals, carbohydrates, unsaturated fatty acids, and protein. It has been reported that the free phenol content and binding phenol in brown rice were increased by 76.67 and 44.64%, respectively after 47 h of germination (Singh et al. 2015; Hithamani and Srinivasan 2014). From the study reviewed, it showed the content of free phenol was higher than the combined phenol. The phenolic acid composition in combined state and free state of brown rice with different germination times is almost similar. However, the phenolic acid composition is significant difference, which is caused by the extraction rate of phenolic acid during germination or hydrolysis of pentanine and polymerization and resynthesis of tannin (Towo et al. 2003; Chethan et al. 2008). Decreased of total phenol contents in grains may also be due to the dissolution of certain polyphenols into the external water environment during germination or through the decomposition and oxidation of polyphenols by stimulated polyphenol esterase and oxidase.

3.2.2.2 Effects of Fermentation on Phytochemicals in Rice

Traditional rice-based fermented foods are well-recognized in many parts of the world including Asia-Pacific countries (Ray et al. 2016). Rice beer or rice-based beverage is a typical food product produces in a few tropical areas, while alcoholic beer made from barley malt in Western countries (Steinkraus 1998). Fermented cereals are well-recognized for their superior digestibility, shelf-life, and nutritional value compared to their unfermented counterparts (Coda et al. 2011). Fermented food has been recognized due to its biological activities and nutritional benefits (Tamang 2015). Fermented cereal-based foods are rich in a wide variety of microbes (from starter culture or natural), in which they are organoleptically and biochemically transformed the substrates, and produce different metabolites and enrich the foods with a range of nutrients such as digestive enzymes, phytochemicals, dietary fibers, fermentable sugars (prebiotic), edible microbes (probiotics), and micronutrients (amino acids, minerals, and vitamins) (Ghosh et al. 2015). Apart from that, the bioactive constituents in rice also possess beneficial effects on the intestinal micro-environments, especially mediating functional behavior and microbial composition. Haria is one of the common traditional rice beer consumed by most of the ethnic from Eastern and Central India (Ghosh et al. 2014). During death feasts, marriage feasts, social feasts, ceremonious occasions, and festival days, sharing of haria by the tribal (adivasi) community is a cultural tradition. The steps of preparation are based on the old empirical knowledge that is transferred by the seniors of the families from one generation to another. During preparation, the rice is boiled to charring and then slowly cools down. Bakhar, a starter culture, is added into boiled rice and stored in an earthen pot for 3–4 days. Upon fermentation, the cream-colored buttermilk like filtrate is consumed with spicy vegetables in the community (Ghosh et al. 2014). The previous study found that wet processing of rice with a culture of the fermentable organisms (*Lactobacillus fermentum* KKL1) enhances the content of phenolics and flavonoids (Ghosh et al. 2015). Such finding indicates the action of acids and microbial enzymes produced by the strain could improve the release of flavonoids and phenolics from their complex form in dietary fiber into a freely soluble form (Katina et al. 2007). Besides phenolics and flavonoids, free radical scavenging activity of fermented rice was increased in a time-dependent manner (Ghosh et al. 2015). This finding is correlated with the finding reported by Kikuzaki et al. (2002), in which the fermented rice extract showed a higher level of flavonoids and free phenolics. In this regard, the isolated organisms play a significant fortifying role in haria fermentation.

Fermentation with rice *koji* has been extensively utilized in the initial stage of manufacturing of fermented foods, for instance, fermented soybean paste (*doenjang* and *miso*), rice wine (*sake* and *makgeolli*), and fermented red pepper paste (*gochujang*) (Blandino et al. 2003; Kum et al. 2015; Shin et al. 2016). Rice *koji* contained enzyme, therefore fermentation with this source may affect the quality of fermented food. Microorganism produces enzymes during rice *koji* fermentation that is involved in the synthesis and hydrolysis of metabolite, thereby improves the bioactivities, taste, and flavor of the fermented food (Kim et al. 2012a; Onuma et al. 2015). Fermentation enhances the bioactive phenolic compounds (Kim et al. 2012b; Liu et al. 2015a). Lee et al. (2016) evaluated the effects of using different fermentation durations and microbial species on the nutritional profiles of rice *koji*. The data revealed that rice *koji* fermented by *Aspergillus oryzae* produces more abundant phenolic acids (ferulic acid and 4-hydroxybenzoic acid) compared to *Bacillus amyloliquefaciens*. However, most of the flavonoids such as tricin, tricin-*O*-glucoside, tricin-7-*O*-rutinoside, and chrysoeriol-rutinoside derived from shikimic acid metabolism were found in *Bacillus amyloliquefaciens* (Lee et al. 2016). Phenolic acids, for instance, ferulic acid are commonly found in the bound form of the cell wall polysaccharides as an integral entity with polysaccharides released by feruloyl esterase (Braga et al. 2014). During rice *koji* fermentation, *Aspergillus oryzae* secretes feruloyl esterase, and thus releases ferulic acid from cell wall polysaccharides (Topakas et al. 2007; Henderson et al. 2012; Braga et al. 2014). In particular, ferulic acid is a component of γ-oryzanol, which is predominantly found in rice bran (Jeng et al. 2012). Furthermore, the impact of fermentation on polyphenols in grains was also presented by Zhai et al. (2015). Zhai et al. (2015) evaluated several grains including sorghum, oat, millet, rice, brown rice, corn, and wheat in solid fermentation of agaricus matsutake and observed that total phenol content (free phenol) of all grains are increased significantly except sorghum after the solid fermentation of agaricus matsutake. Such findings may be linked to the strong metabolism of agaricus matsutake that produces phenolic compounds. Collectively, the effects of chemical and physical treatments on polyphenols during rice processing need to be further elucidated. More effective treatment methods also need to be evaluated to preserve nutritional value while improving flavor and taste.

3.3 Rice By-products

The combination of rice processing steps produces several rice by-products such as rice husk, rice bran, broken rice, and rice straw (Peanparkdee and Iwamoto 2019). Each tonne of harvested paddy rice can produce 1.35 tonnes of rice straw in the field (Moraes et al. 2014). Notably, each tonne of processed paddy rice produces 140 kg of broken rice, 100 kg of rice bran, and 200 kg of rice husk. Indeed, the major rice by-product is husk and on average it represents 20% of the paddy produced (Tong et al. 2018). Broken rice and rice bran can be produced in different ratios, depending on its efficiency and rice milling rate (Muthayya et al. 2014). A study reported by Lim et al. (2013) showed that one of the major challenges facing by the rice industry is the disposal or appropriate use of rice by-products, particularly in underdeveloped countries. In general, the sales of rice by-products are a convenient option to decrease environmental problems and obtain more income. Rice by-products are usually used as animal feed (Sharif et al. 2014). Nonetheless, it would be more appropriate and profitable to invest in technology to convert it into value-added products.

3.3.1 Rice Bran

Paddy or rough rice consists of white starchy rice kernel attached with a coating of bran, enclosed with a tough siliceous husk (Eyarkai Nambi et al. 2017). The bran layer is exposed to air when the husk is removed, and thus developed off-flavor in brown rice due to its endogenous lipase. Additionally, the appearance of brown rice is not attractive due to its color. Therefore, further processing of rice is needed by removing bran from brown rice to produce white rice (Ding et al. 2018), but the consumption of germinated brown rice or brown rice is getting popular in recent decades due to their potential health benefits (Mohan et al. 2017; Ravichanthiran et al. 2018). Rice bran constitutes nearly 10% of the weight of paddy (Zhao et al. 2018), comprising of subaleurone, aleurone, testa, inner pericarp, as well as the embryo and/or some portion of broken rice (Pandey and Shrivastava 2018). The percentage of rice bran is highly dependent on the rate of milling system and the type of rice (Sharif et al. 2014). In traditional, rice bran is usually considered as a waste and is either discarded or used as an animal feed (Patel 2007; Khan et al. 2009). In China, nearly 60–70% of rice bran oil (RBO) production is inedible, and thus rice bran is usually utilized as cattle food (Moraes et al. 2014).

In general, the macro- and micronutrient composition of rice bran varies based on the post-harvesting treatments, rice variety, climatologic condition, degree of milling, and the soil (Goufo and Trindade 2017). During the rice milling process, crude rice bran is produced. However, several processes need to be performed before it can be used. Rice bran is sweet in taste, slightly toasted nutty flavor, moderately oily, and light in color (Lavanya et al. 2017). A high content of free fatty acid led to the development of rancidity. Accordingly, the processing of rice bran was performed to inactivate nutritional inhibitors and lipases without destroying the protein quality of rice bran (Irakli et al. 2018). In addition, it can also destroy the insect, bacteria, and fungi infestation, and therefore enhance the shelf-life. However, only a small fraction of rice bran can be used to produce commercial food products after stabilization, yet most of the rice bran is used as a fuel in boilers or as an ingredient in animal feed (Nunoi et al. 2019). Upon rice milling processing, the oil is exposed to lipases (Kim et al. 2012b; Gul et al. 2015) and thus breaks down to free fatty acids at 5–7% of the weight of oil per day. This unfavorable effect could be attributed to instability during storage (Guevara-Guerrero et al. 2019).

Despite rice bran exerts nutritional proteins and natural antioxidants (Jariwalla 2001; Fabian and Ju 2011), its potential use as a functional food is limited by its insolubility of protein and the integrity of its nutraceutical compounds, primarily refers to the phenolic compounds. Subsequently, the therapeutic potential of oil derived from rice bran has been evaluated in recent decades. A study reported by Amarasinghe and Gangodavilage (2004) evaluated the oil extracted from different types of rice bran available in Sri Lanka using different methods and parameters. The data showed that rice bran consists of 20% of oil, predominantly from linoleic, oleic, and palmitic acids. Free fatty acids in oil extracted from rice bran are relatively higher than other edible oils due to the lipase activity (Yi and Kim 2019). Subsequently, this converts the RBO into the free fatty acids and glycerol which gives the product a bitter taste and rancid smell that renders the bran unsuitable for consumption (Arora et al. 2015). Due to the naturally occurring hydrolytic rancidity and enzymatic activity, the stabilization is vitally important to control undesirable conditions (Wang et al. 2017a). Thermal and chemical treatments are the most common stabilization procedures (So et al. 2018; Liu et al. 2019). Among the heat stabilization, microwave heating is an effective technique for stabilization of rice bran (Lavanya et al. 2019). Rice bran was stabilized using a microwave by heating 4 min for 110–115 °C to denature the enzymes (Zhu 2002). After stabilization, the bran goes through further processing before the oil can be consumed (Rafe et al. 2017). On average, rice bran comprised of 10–23% of RBO (Kennedy and Burlingame 2003; Friedman 2013). RBO can be extracted using three common methods such as solvent extraction, X-M milling, and hydraulic pressing (Nagendra Prasad et al. 2011). Among these methods, solvent extraction is one of the most common methods for oil extraction (Yanık 2017). Solvent extraction is preferable compared to other methods and this could be attributed to the extraction yields (0.549 g RBO/g) and the solvent is easily removed via purification procedures (Chiou et al. 2013). Although hexane has been used as the commercial extraction solvents, it exerts some limitations as it can affect the oil color and considered hazardous (Nagendra Prasad et al. 2011). There are two by-products produced during extraction, namely defatted rice bran and crude bran oil (Bessa et al. 2017). Crude RBO contains 90% lipids, 4% free fatty acids, and 4% unsaponifiable (oil, fat, and wax). The crude RBO is refined by removing the free fatty acids and thus improves sensory properties and reduces the rancidity (Khan et al. 2011; Vaisali et al. 2015).

3.3.2 Rice Germ

In addition to rice bran, rice germ or known as an embryo or reproductive portions is another by-product produced from rice processing which grows and germinates into a plant (Organization for Economic Co-Operation and Development 2004). The embryo is relatively small and located at the ventral side in the base of the grain. In particular, it is enclosed with fibrous cellular remains in the nucellus, seed coat, and pericarp and covered with a single aleurone layer (Bechtel and Pomeranz 1977). The embryo consists of the embryonic axis and scutellum (cotyledon). The scutellum is rich in globoid particles resembling aleurone grains (Tanaka et al. 1977). Rice germ can be obtained inexpensively, is an agricultural by-product produced during rice processing (Tan and Norhaizan 2017). The germ is usually removed with the bran. In this regard, rice germ is usually collected with the bran during the milling process. When separated, it has the highest nutritional value among all the rice by-products (Kik 1952).

3.3.3 Rice Straw

Rice straw is removed during harvesting of the rice grain. It is approximately 50% of the dry weight of rice, varied from 40–60% according to the harvesting methods, field condition, and methods of cultivation (Kadam et al. 2000). Each tonne of rice grain harvested can produce 1.35 tonnes of rice straw (Kadam et al. 2000). Rice straw is a fibrous lignocellulosic material, which is distinguished from other residues due to its high amounts of silicon dioxide, known as silica (SiO_2). The ash content varies from 13 to 20% depends on the condition after harvesting. Typically, ash comprised of 1.3% CaO, 3% Fe_2O_3, 3% P_2O_5, 10% K_2O, and 75% SiO_2 with a small fraction of Na and Mg (Kadam et al. 2000). The composition of rice straw is varied from plant species, regions, and seasons (Binod et al. 2010).

In general, rice straw is burned in the open field, in which this is a low cost of disposal method and may prevent the proliferation of fungi in the field (Kadam et al. 2000). For instance, nearly 23% of rice straw in India is burned in the open or left in the field. Approximately 95% is burned in the open field in the Philippines and 48% in Thailand (Pandey et al. 2010). Despite open-field burning is the easiest way; it produces a huge level of greenhouse gases, for example, particulate matter, nitric oxide (NO), sulfur oxide, and carbon oxide (Romasanta et al. 2017). Besides air pollution, burning straw can cause traffic accidents when the field is near to the roads, owing to the generation of large amounts of smoke (Pham et al. 2019).

Several studies have demonstrated that rice straw can be a valuable energy resource, rather than being wasted by open burning. Rice straw has the potential to be used as a source of biomass for power generation. Biomass is a source of sustainable energy to neutralize carbon dioxide. Rice straw has been recognized as a biofuel for energy and steam generation in recent decades (Abraham et al. 2016). It has shown great interest as an alternative to fossil fuels (Hu et al. 2019). A study by Gadde et al. (2009) focusing on the amount of rice straw produced and the energy potential in the Philippines, Thailand, and India. The data revealed that the energetic potential of rice straw as a renewable fuel is 142, 237, and 312 × 10^{15} J in the Philippines, Thailand, and India, respectively (Gadde et al. 2009). Nonetheless, the energy produced directly from rice straw is complex; depending on the several factors, for instance, low density, low ash melting point, and high ash content (Singh et al. 2016). Data reported by Yang et al. (2016) have shown the feasibility of developing a solid biomass fuel from rice straw.

3.3.4 Rice Husk

Rice husk or rice hull is an abundant agriculture waste in many rice producing countries (Kumar et al. 2013). It is a protective or coating layer formed during grain growth, with high volume and low density. The caryopsis is enclosed by the husk, comprised of two "modified" leaves (lemmae), namely larger lemma (ventral) and palea (dorsal) (Juliano and Tuaño 2019). The tightness of the husk and the ability of the palea and lemma to hook together without gaps have been linked to grain resistance to insect infestation during storage (Juliano and Tuaño 2019). Rice husk is equivalent to nearly 20% of the grain weight and comprised of four structural, spongy, fibrous or cell layers. The primary constituents of rice husk are inorganic residues (20%), lignin (30%), and cellulose (50%). The inorganic residue consists of 95–98% by weight of silica in the amorphous hydrated form (Javed et al. 2010). High levels of silica and lignin enable the husk to show an antioxidative defense system that prevents the seeds from oxidative stress (Kim et al. 2012b). Rice husk has a packing density of 122 kg/m^3 and a dimension of 2–3 mm wide, 0.2 mm thick, and 8–10 mm long (Fang et al. 2004). The shells are low-cost raw materials that are hard to reuse due to their large volume, resistance to degradation, low nutritional properties, and abrasion (Calheiro 2011). Rice husk is inedible due to high silica content and low nutritional properties (Madandoust et al. 2011; Roschat et al. 2016), which is used predominantly in many non-food applications as an agricultural waste product. The silica provides a protective feature to the husk and allows the rice husk to burn slowly (Nazari et al. 2019). Accordingly, several forms of silica can be produced during ash formation controlled by different temperatures during burning (Singh et al. 2008).

Rice husk is predominantly used for power generation (Quispe et al. 2017). Using rice husk for power generation could decrease the environmental problems caused by the disposal of waste, and thereby leading to the use of renewable sources and reduce dependence on petroleum (Unrean et al. 2018). For example, rice husk has been utilized as a fuel to generate power and steam in southern Brazil (Nadaleti 2019). Among 484 plants operating from biomass burning in Brazil, nine are powered exclusively using rice husks with an installed capacity to produce 36.4 GW (ANEEL—Agência Nacional de Energia Elétrica 2014). Furthermore, other potential sources of energy such as hydrogen and methane can be generated through biomass conversion processes. Besides hydrogen and methane, rice husk can also produce ethanol (Omidvar et al. 2016). Rice husk is a promising material to be used as a poultry litter because it could enhance weight gain for the animal (Gao et al. 2015).

Rice husk ash is the common term used to describe all types of ash generated when burning rice husk (Xiong et al. 2009). In general, the ash produced is varied considerably (about 17–26%), depending on the burning condition of the husk (Bie et al. 2015) such as the burning process applied (fluidized bed or grate), type of equipment, temperature, and time (Asavapisit and Ruengrit 2005; Hwang et al. 2011). The amorphous ash is produced when the rice husk is burnt at 700 °C. At the temperature higher than 850 °C, the husk produces crystalline ash (Singh et al. 2008; Chao-Lung et al. 2011). Disposing rice husk ash directly into the environment could bring a negative impact on the environment (Khan et al. 2012). The detrimental impact could be attributed to the presence of high silica content and residual carbon, in which it can cause a detrimental effect, for instance, accumulation of ash in river beds and soil acidification (Makul 2019).

Rice husk ash is highly porous, bulky, and lightweight material with a density of nearly 1800 kg/m^3 (Kumar et al. 2013). In general, rice husk ash contains CaO, P_2O_5, K_2O, C, and SiO_2 with minor amounts of Na, Fe, and Mg (Suryana et al. 2018). The previous finding has reported that Zn, Mn, Cu, Fe, Mg, Ca, K, and Na are mostly found in ashes, and differences in the composition could be attributed to analysis methods, sample preparation, year of harvest, and geographical factors (Wang et al. 2017). Nehdi et al. (2003) and Armesto et al. (2002) further found that rice husk ash obtained from a fluidized bed reactor showed high amounts of silica and low values of ignition and carbon loss compared to other burning conditions. This finding implies that ashes can be obtained from burning without temperature control (Calheiro 2011; Zain et al. 2011), and have higher levels of ignition loss compared to controlled burning (Vayghan et al. 2013). Rice husk ash is inert in characteristics, which makes it favorably to transform into a by-product, and therefore safe for reuse and recycling. One of the prominent uses of rice husk is the conversion into fuel (Hossain et al. 2018). When husk separated from the rice kernel, it could be used as a source of energy due to the organic compounds (International Rice Research Institute (IRRI) 2016). For example, countries with high production of rice, for instance, India process the husk into the fuel. The energy can be produced from thermal processing such as biofuels, syngas, electric, and heat (Pradhan et al. 2013). An emerging study has shown the potential of rice husk ash in the preparation of materials for different industries, for instance, construction, refractory aggregates, refractories, pressed bricks, thermal insulation, and glassmaking (Della et al. 2001; Chiang et al. 2009; Ewais et al. 2017).

3.3.5 Broken Rice

After undergo rice polishing process, the rice is graded based on the size and for those do not meet the required size is classified as broken (Bodie et al. 2019). The size of broken rice is less than three-fourths of the whole kernel (Van Dalen 2004). The rice grain undergoes mechanical tensions during processing can break the grains and thus generate broken rice. Broken rice is also known as rice grits, consists of defective and broken grains (Bassinello et al. 2015). About 14% of broken rice is produced during the rice milling process. According to the United States Department of Agriculture (USDA) (2009), broken rice can be classified into three groups, namely brewers' rice, screenings, and heads. These names represent small, intermediate, and large fragments, respectively (United States Department of Agriculture 2009). In general, the broken rice can be mixed with the unbroken rice in certain composition, based on the legislation of the countries. Notably, broken rice has some opportunities for recycling, for example, production of beer, animal feeding, and production of rice flour (Mukhopadhyay and Siebenmorgen 2017). The protein composition in broken rice is 6–8%. Although broken rice has low amounts of protein, this protein is regarded as very valuable because it is hypoallergenic and safe for human consumption (Moraes et al. 2014). A previous study has shown that nearly 10% of the total rice consumed in the US is utilized as the brewery industry and pet food (United States Department of Agriculture 2016). Small mills produce 25% broken kernels and this can be as high as 10–15% in large and medium rice mills and depends on insect infection, relative humidity, immature concerns, chalkiness, and moisture absorption (Siebenmorgen et al. 1998; Muthayya et al. 2014; Bruce and Atungulu 2018).

The previous study suggested that broken rice can be used as an ingredient for the fermentation of ethanol. A study reported by Li et al. (2014) evaluated the utilization of broken rice for the production of Chinese rice wine. The data revealed that the extrusion enzyme enhanced the production of alcohol and the proliferation of yeast during the fermentation process. From the study reviewed, it showed that the efficiency and fermentation of processed rice wine were 94.66 and 38.07%, respectively. In addition, broken rice can serve as a crucial raw ingredient for the production of lactic acid and other functional material. For instance, lactic acid is produced from broken rice using *Lactobacillus delbrueckii* (Nakano et al. 2012).

3.3.6 Brewers' Rice

In addition to broken rice, rice bran, rice straw, rice husk, and rice germ, brewers' rice is another by-product from the rice industry. Brewers' rice contained a mixture of rice bran and rice germ with broken kernels (Nordin et al. 2014). It is the smallest and the last milling fraction that is removed during the milling of paddy rice and is often separated from larger rice kernels (Association of American feed Control

Officials 2011). The size of brewers' rice is less than one-quarter of the full kernel of milled rice. Typically, brewers' rice is utilized as a brewing ingredient and animal feed (Glatthar et al. 2003).

References

Abraham A, Mathew AK, Sindhu R et al (2016) Potential of rice straw for bio-refning: an overview. Bioresour Technol 215:29–36

Ahmad Hanis IAH, Jinap S, Mad Nasir S et al (2012) Consumers' demand and willingness to pay for rice attributes in Malaysia. Int Food Res J 19:363–369

Ajala AS, Gana A (2015) Analysis of challenges facing rice processing in Nigeria. J Food Process 2015:893673. 6p

Amarasinghe BMWPK, Gangodavilage NC (2004) Rice bran oil extraction in Sri Lanka data for process equipment design. Food Bioprod Process 82:54–59

Andrés S, Jaramillo E, Bodas R et al (2018) Grain grinding size of cereals in complete pelleted diets for growing lambs: effects on ruminal microbiota and fermentation. Small Rumin Res 159:38–44

ANEEL—Agência Nacional de Energia Elétrica (2014) Matriz Energética do Brasil. http://www.aneel.gov.br/aplicacoes/capacidadebrasil/OperacaoGeracaoTipo.asp?tipo=5&ger=Combustivel&principal=Biomassa. Accessed 7 Aug 2014

Anunciação PC, de Morais CL, de Cássia Gonçalves Alfenas R et al (2019) Extruded sorghum consumption associated with a caloric restricted diet reduces body fat in overweight men: a randomized controlled trial. Food Res Int 119:693–700

Arianti R, Oktarina O (2018) Benefit of purple Aruk rice (Siangu) in lowering body mass index (BMI) and body fat percentage. Scientiae Educatia Jurnal Pendidikan Sains 7:21–32

Armesto L, Bahill A, Veijonen K et al (2002) Combustion behavior of rice husk in a bubbing fluidised bed. Biomass Bioenergy 23:171–179

Arora R, Toor AP, Wanchoo RK (2015) Esterification of high free fatty acid rice bran oil: parametric and kinetic study. Chem Biochem Eng 29:617–623

Arribas C, Cabellos B, Sánchez C et al (2017) The impact of extrusion on the nutritional composition, dietary fiber and in vitro digestibility of gluten-free snacks based on rice, pea and carob flour blends. Food Funct 8:3654–3663

Asavapisit S, Ruengrit N (2005) The role of RHA-blended cement in stabilizing metal-containing wastes. Cem Concr Compos 27:782–787

Association of American feed Control Officials (2011) Brewers rice. http://www.aafco.org/

Baccaro S, Bal O, Cemmi A et al (2018) The effect of gamma irradiation on rice protein aqueous solution. Radiat Phys Chem 146:1–4

Bassinello PZ, Carvalho AV, de Oliveira RA et al (2015) Expanded gluten-free extrudates made from rice grits and Bandinha (bean) flour mixes: main quality properties. J Food Process Preserv 39:2267–2275

Bechtel DB, Pomeranz Y (1977) Ultrastructure of the mature ungerminated rice (*Oryza sativa*) caroypsis. The caryopsis coat and aleurone cells. Am J Bot 64:966–973

Bessa LCBA, Ferreira MC, Rodrigues CEC et al (2017) Simulation and process design of continuous countercurrent ethanolic extraction of rice bran oil. J Food Eng 202:99–113

Bie R-S, Song X-F, Liu Q-Q et al (2015) Studies on effects of burning conditions and rice husk ash (RHA) blending amount on the mechanical behavior of cement. Cem Concr Compos 55:162–168

Binod P, Sindhu R, Singhania RR et al (2010) Bioethanol production from rice straw: an overview. Bioresour Technol 10:4767–4774

Blandino A, Al-Aseeri ME, Pandiella SS et al (2003) Cereal-based fermented foods and beverages. Food Res Int 36:527–543
Bodie AR, Micciche AC, Atungulu GG et al (2019) Current trends of rice milling byproducts for agricultural applications and alternative food production systems. Front Sustain Food Syst 3:47
Bolea CA, Grigore-Gurgu L, Aprodu L et al (2019) Process-structure-function in association with the main bioactive of black rice flour sieving fractions. Foods 8:131
Borah A, Mahanta CL, Kalita D (2016) Optimization of process parameters for extrusion cooking of low amylose rice flour blended with seeded banana and carambola pomace for development of minerals and fber rich breakfast cereal. J Food Sci Technol 53:221–232
Braga CMP, Delabona Pda S, Lima DJ et al (2014) Addition of feruloyl esterase and xylanase produced on-site improves sugarcane bagasse hydrolysis. Bioresour Technol 170:316–324
Bruce RM, Atungulu GG (2018) Assessment of pasting characteristics of size fractionated industrial parboiled and non-parboiled broken rice. Cereal Chem 95:889–899
Calheiro D (2011) Influência do uso de aditivos na moagem de cinzas de casca de arroz para sua adequação como coproduto. Master Dissertation in Civil Engineering, University of Vale do Rio dos Sinos, RS, Brazil
Cardoso LM, Montini TA, Pinheiro SS et al (2014) Effects of processing with dry heat and wet heat on the antioxidant profile of sorghum. Food Chem 152:210–217
Chalermchaiwat P, Jangchud K, Jangchud A et al (2015) Antioxidant activity, free gamma-aminobutyric acid content, selected physical properties and consumer acceptance of germinated brown rice extrudates as affected by extrusion process. LWT Food Sci Technol 64:490–496
Chanvrier H, Nordstrom Pillin C, Vandeputte G et al (2015) Impact of extrusion parameters on the properties of rice products: a physicochemical and X-ray tomography study. Food Struct 6:29–40
Chao-Lung H, Le Anh-Tuan B, Chun-Tsun C (2011) Effect of rice husk ash on the strength and durability characteristics of concrete. Constr Build Mater 25:3768–3772
Chen H, Liu RH (2018) Potential mechanisms of action of dietary phytochemicals for cancer prevention by targeting cellular signaling transduction pathways. J Agric Food Chem 66:3260–3276
Chen MH, McClung AM, Bergman CJ (2016) Bran data of total flavonoid and total phenolic contents, oxygen radical absorbance capacity, and profiles of proanthocyanidins and whole grain physical traits of 32 red and purple rice varieties. Data Brief 8:6–13
Chethan S, Sreerama YN, Malleshi NG (2008) Mode of inhibition of finger millet malt amylases by the milletphenolics. Food Chem 111:187–191
Chiang K-Y, Chou P-H, Hua C-R et al (2009) Lightweight bricks manufactured from water treatment sludge and rice husks. J Hazard Mater 171:76–82
Chinnaswamy R, Hanna MA (1990) Macromolecular and functional properties of native and extrusion-cooked corn starch. Cereal Chem 67:490–499
Chinprahast N, Tungsomboon T, Nagao P (2016) Antioxidant activities of Thai pigmented rice cultivars and application in sunflower oil. Int J Food Sci Technol 51:46–53
Chiou TY, Ogino A, Kobayashi T et al (2013) Characteristics and antioxidative ability of defatted rice bran extracts obtained using several extractants under subcritical conditions. J Oleo Sci 62:1–8
Coda R, Rizzello CG, Trani A et al (2011) Manufacture and characterization of functional emmer beverages fermented by selected lactic acid bacteria. Food Microbiol 28:526–536
Della VP, Kühn I, Hotza D (2001) Caracterização de cinza de casca de arroz para uso como matéria-prima na fabricação de refratários de silica [Characterization of rice husk ash for use as raw material in the manufacture of silica refractory]. Química Nova 24:778–782
Dhankhar P (2014) Rice milling. IOSR J Eng 4:34–42
Ding C, Khir R, Pan Z et al (2018) Influence of infrared drying on storage characteristics of brown rice. Food Chem 264:149–156
Ding C, Liu Q, Li P et al (2019) Distribution and quantitative analysis of phenolic compounds in fractions of *Japonica* and *Indica* rice. Food Chem 274:384–391

dos Santos PA, Caliari M, Júnior MSS et al (2019) Use of agricultural by-products in extruded gluten-free breakfast cereals. Food Chem 297:124956
Ewais EMM, Elsaadany RM, Ahmed AA et al (2017) Insulating refractory bricks from water treatment sludge and rice husk ash. Refract Ind Ceram 58:136–144
Eyarkai Nambi V, Manickavasagan A, Shahir S (2017) Rice milling technology to produce brown rice. In: Manickavasagan A, Santhakumar C, Venkatachalapathy N (eds) Brown rice. Springer, Cham, pp 3–21
Fabian C, Ju YH (2011) A review on rice bran protein: its properties and extraction methods. Crit Rev Food Sci Nutr 51:816–827
Fang M, Yang L, Chen G et al (2004) Experimental study on rice husk combustion in a circulating fluidized bed. Fuel Process Technol 85:1273–1282
Finocchiaro F, Ferrari B, Gianinetti A et al (2007) Characterization of antioxidant compounds ofred and white rice and changes in total antioxidantcapacity during processing. Mol Nutr Food Res 51:1006–1019
Friedman W (2013) Rice brans, rice bran oils, and rice hulls: composition, food and industrial uses, and bioactivities in humans, animals, and cells. J Agric Food Chem 61:10626–10641
Gadde B, Menke C, Wassmann R (2009) Rice straw as a renewable energy source in India, Thailand, and the Philippines: overall potential and limitations for energy contribution and greenhouse gas mitigation. Biomass Bioenergy 33:1532–1546
Gao H, Zhou C, Wang R et al (2015) Comparison and evaluation of co-composting corn stalk or rice husk with swine waste in China. Waste Biomass Valoriz 6:699–710
Garzón AG, Torres RL, Drago SR (2019) Changes in phenolics, γ-aminobutyric acid content and antioxidant, antihypertensive and hypoglycaemic properties during ale white sorghum (Sorghum bicolor (L.) Moench) brewing process. Int J Food Sci Technol 54:1901–1908
Ghosh K, Maity C, Adak A et al (2014) Ethnic preparation of haria, a rice-based fermented beverage, in the province of lateritic West Bengal, India. Ethnobot Res Appl 12:39–49
Ghosh K, Ray M, Adak A et al (2015) Role of probiotic *Lactobacillus fermentum* KKL1 in the preparation of a rice based fermented beverage. Bioresour Technol 188:161–168
Glatthar J, Heinisch J, Senn T (2003) The use of unmalted triticale in brewing and its effect on wort and beer quality. J Am Soc Brew Chem 61:182–190
Gong ES, Luo SJ, Li T et al (2017) Phytochemical profiles and antioxidant activity of brown rice varieties. Food Chem 227:432–443
Goufo P, Trindade H (2017) Factors influencing antioxidant compounds in rice. Crit Rev Food Sci Nutr 57:893–922
Guevara-Guerrero B, Fernández-Quintero A, Montero-Montero JC (2019) Free fatty acids in rice bran during its storage after a treatment by twin-screw extrusion to prevent possible rapid hydrolytic rancidity of lipids. DYNA 86:177–181
Gujral HS, Sharma P, Kumar A et al (2012) Total phenolic content and antioxidant activity of extruded brown rice. Int J Food Prop 15:301–311
Gul K, Yousuf B, Singh AK et al (2015) A rice bran: nutritional values and its emerging potential for development of functional food—a review. Bioact Carbohydr Diet Fibre 6:24–30
Hagenimana A, Ding X, Fang T (2006) Evaluation of rice flour modified by extrusion cooking. J Cereal Sci 43:38–46
Han SW, Chee KM, Cho SJ (2015) Nutritional quality of rice bran protein in comparison to animal and vegetable protein. Food Chem 172:766–769
Henderson AJ, Ollila CA, Kumar A et al (2012) Chemopreventive properties of dietary rice bran: current status and future prospects. Adv Nutr 3:643–653
Heo S, Lee SM, Shim J-H et al (2013) Effect of dry- and wet-milled rice flours on the quality attributes of gluten-free dough and noodles. J Food Eng 116:213–217
Hithamani G, Srinivasan K (2014) Bioaccessibility of polyphenols from wheat (Triticum aestivum), sorghum (Sorghum bicolor), green gram (Vigna radiata) and chickpea (Cicer arietinum) as influenced by domestic food processing. J Agric Food Chem 62:11170–11179

Horstmann SW, Lynch KM, Arendt EK (2017) Starch characteristics linked to gluten-free products. Foods 6:29

Hossain SS, Mathur L, Roy PK (2018) Rice husk/rice husk ash as an alternative source of silica in ceramics: a review. J Asian Ceramic Soc 6:299–313

Hu Z, Tang X, Zhang M et al (2018) Effects of different extrusion temperatures on extrusion behavior, phenolic acids, antioxidant activity, anthocyanins and phytosterols of black rice. RSC Adv 8:7123–7132

Hu J, Li C, Zhang Q et al (2019) Using chemical looping gasification with Fe_2O_3/Al_2O_3 oxygen carrier to produce syngas (H_2+CO) from rice straw. Int J Hydrog Energy 44:3382–3386

Huang Y-P, Lai H-M (2016) Bioactive compounds and antioxidative activity of colored rice bran. J Food Drug Anal 24:564–574

Hwang C-L, Bui LA-T, Chen C-T (2011) Effect of rice husk ash on the strength and durability characteristics of concrete. Construct Build Mater 25:3768–3772

International Rice Research Institute (IRRI) (2016) Milling byproducts. Manila: International Rice Research Institute (IRRI). http://www.knowledgebank. Accessed 21 Mar 2019

Irakli MN, Katsantonis DN, Ward S (2017) Rice bran: a promising natural antioxidant component in breadmaking. J Nutr Food Sci 2:14

Irakli M, Kleisiaris F, Mygdalia A et al (2018) Stabilization of rice bran and its effect on bioactive compounds content, antioxidant activity and storage stability during infrared radiation heating. J Cereal Sci 80:135–142

Jariwalla RJ (2001) Rice-bran products: phytonutrients with potential applications in preventive and clinical medicine. Drugs Exp Clin Res 27:17–26

Javed SH, Naveed S, Ramzan N (2010) Characterization of amorphous silica obtained from KMnO/sub 4/treated rice husk. J Chem Soc Pak 32:78–82

Jeng TL, Shih YJ, Ho PT et al (2012) γ-oryzanol, tocol and mineral compositions in different grain fractions of giant embryo rice mutants. J Sci Food Agric 92:1468–1474

Juliano BO, Tuaño APP (2019) Chapter 2 Gross structure and composition of the rice grain. In: Rice. 4th edn. Chemistry and Technology, pp 31–53

Kadam KL, Forrest LH, Jacobson WA (2000) Rice straw as a lignocellulosic resource: collection, processing, transportation, and environmental aspects. Biomass Bioenergy 18:369–389

Katina K, Laitila A, Juvonen R et al (2007) Bran fermentation as a means to enhance technological properties and bioactivity of rye. Food Microbiol 24:175–186

Kennedy G, Burlingame B (2003) Analysis of food composition data on rice from a plant genetic resources perspective. Food Chem 80:589–596

Khan MAI, Ueno K, Horimoto S et al (2009) Physicochemical, including spectroscopic, and biological analyses during composting of green tea waste and rice bran. Biol Fertil Soils 45:305–313

Khan SH, Butt MS, Sharif MK et al (2011) Functional properties of protein isolates extracted from stabilized rice bran by microwave, dry heat, and parboiling. J Agric Food Chem 59:2416–2420

Khan R, Jabbar A, Ahmad I et al (2012) Reduction in environmental problems using rice-husk ash in concrete. Construct Build Mater 30:360–365

Kik MC (1952) The nutritive value of rice and its byproducts. Ark Agr Exp Station Bulletin No. 589, Fayetteville

Kikuzaki H, Hisamoto M, Hirose K et al (2002) Antioxidant properties of ferulic acid and its related compounds. J Agric Food Chem 50:2161–2168

Kim AJ, Choi JN, Kim J et al (2012a) Metabolomics-based optimal koji fermentation for tyrosinase inhibition supplemented with Astragalus Radix. Biosci Biotechnol Biochem 76:863–869

Kim HY, Hwang IG, Kim TM et al (2012b) Chemical and functional components in different parts of rough rice (*Oryza sativa* L.) before and after germination. Food Chem 134:288–293

Kovacic P, Somanathan R (2011) Cell signaling and receptors with resorcinols and flavonoids: redox, reactive oxygen species, and physiological effects. J Recept Signal Transduct Res 31:265–270

Kum SJ, Yang SO, Lee SM et al (2015) Effects of *Aspergillus* species inoculation and their enzymatic activities on the formation of volatile components in fermented soybean paste (doenjang). J Agric Food Chem 63:1401–1418
Kumar S, Sangwan P, Dhankhar RMV et al (2013) Utilization of rice husk and their ash: a review. Res J Chem Environ Sci 1:126–129
Lavanya MN, Venkatachalapathy N, Manickavasagan A (2017) Physicochemical characteristics of rice bran. In: Manickavasagan A, Santhakumar C, Venkatachalapathy N (eds) Brown rice. Springer, Cham
Lavanya MN, Saikiran KCHS, Venkatachalapathy N (2019) Stabilization of rice bran milling fractions using microwave heating and its effect on storage. J Food Sci Technol 56:889–895
Lee DE, Lee S, Jang ES et al (2016) Metabolomic profiles of *Aspergillus oryzae* and *Bacillus amyloliquefaciens* during rice *koji* fermentation. Molecules 21:773
Li H, Wei B, Wu C et al (2014) Modelling and optimisation of enzymatic extrusion pretreatment of broken rice for rice wine manufacture. Food Chem 150:94–98
Lim JS, Manan ZA, Hashim H et al (2013) Towards an integrated, resource-efficient rice mill complex. Resour Conserv Recycl 75:41–51
Ling WH, Cheng QX, Ma J et al (2001) Red and black rice decrease atherosclerotic plaque formation and increase antioxidant status in rabbits. J Nutr 131:1421–1426
Liu L, Guo J, Zhang R et al (2015a) Effect of degree of milling on phenolic profiles and cellular antioxidant activity of whole brown rice. Food Chem 185:318–325
Liu SX, Yang HY, Li SY et al (2015b) Polyphenolic compositions and chromatic characteristics of bog bilberry syrup wines. Molecules 20:19865–19877
Liu K-L, Zheng J-b, F-s C (2017) Relationships between degree of milling and loss of Vitamin B, minerals, and change in amino acid composition of brown rice. LWT Food Sci Technol 82:429–436
Liu K, Zhao S, Li Y et al (2018) Analysis of volatiles in brown rice, germinated brown rice, and selenised germinated brown rice during storage at different vacuum levels. J Sci Food Agric 98:2295–2301
Liu YQ, Strappe P, Zhou ZK et al (2019) Impact on the nutritional attributes of rice bran following various stabilization procedures. Crit Rev Food Sci Nutr 59:2458–2466
Lohani UC, Muthukumarappan K (2017) Effect of extrusion processing parameters on antioxidant, textural and functional properties of hydrodynamic cavitated corn flour, sorghum flour and apple pomace-based extrudates. J Food Process Eng 40:e12424
Luh BS, Mickus RR (1991) Parboiled rice. In: Luh BS (ed) Rice utilization, 2nd edn. Van Nostrand Reinhold, New York, pp 51–88
Madandoust R, Ranjbar MM, Moghadam HA et al (2011) Mechanical properties and durability assessment of rice husk ash concrete. Biosyst Eng 110:144–152
Makul N (2019) Combined use of untreated-waste rice husk ash and foundry sand waste in high-performance self-consolidating concrete. Res Mater 1:100014
Milovanovic V, Smutka L (2017) Asian countries in the global rice market. Acta Univ Agric Silvic Mendelianae Brun 65:679–688
Mishra A, Mishra HN, Srinivasa Rao P (2012) Preparation of rice analogues using extrusion technology. Int J Food Sci Technol 47:1789–1797
Mohan V, Ruchi V, Gayathri R et al (2017) Hurdles in brown rice consumption. In: Manickavasagan A, Santhakumar C, Venkatachalapathy N (eds) Brown rice. Springer, Cham
Moraes CAM, Fernandes LJ, Calheiro D et al (2014) Review of the rice production cycle: byproducts and the main applications focusing on rice husk combustion and ash recycling. Waste Manag Res 32:1034–1048
Mukhopadhyay S, Siebenmorgen TJ (2017) Physical and functional characteristics of broken rice kernels caused by moisture-adsorption fissuring. Cereal Chem 94:539–545
Muthayya S, Sugimoto JD, Montgomery S et al (2014) An overview of global rice production, supply, trade, and consumption. Ann N Y Acad Sci 1324:7–14

Nadaleti WC (2019) Utilization of residues from rice parboiling industries in southern Brazil for biogas and hydrogen-syngas generation: heat, electricity and energy planning. Renew Energy 131:55–72

Nagendra Prasad MN, Sanjay KR, Shravya Khatokar M et al (2011) Health benefits of rice bran–a review. J Nutr Food Sci 1:1–7

Nakano S, Ugwu CU, Tokiwa Y (2012) Efficient production of D-(-)-lactic acid from broken rice by Lactobacillus delbrueckii using Ca(OH)2 as a neutralizing agent. Bioresour Technol 104:791–794

Nascimento TA, Calado V, Carvalho CW (2017) Effect of Brewer's spent grain and temperature on physical properties of expanded extrudates from rice. LWT Food Sci Technol 79:145–151

Nazari MM, San CP, Atan NA (2019) Combustion performance of biomass composite briquette from rice husk and banana residue. Int J Adv Sci Eng Inform Technol 9:455–460

Neder-Suárez D, Amaya-Guerra CA, Báez-González et al (2018) Resistant starch formation from corn starch by combining acid hydrolysis with extrusion cooking and hydrothermal storage. Starch 70:1700118

Nehdi M, Duquette J, El Damatty A (2003) Performance of rice husk ash produced using a new technology as mineral admixtures in concrete. Cem Concr Res 33:1203–1210

Nordin NNAM, Karim R, Ghazali HM et al (2014) Effects of various stabilization techniques on the nutritional quality and antioxidant potential of brewer's rice. J Eng Sci Technol 9:347–363

Nunoi A, Wanapat M, Foiklang S et al (2019) Effects of replacing rice bran with tamarind seed meal in concentrate mixture diets on the changes in ruminal ecology and feed utilization of dairy steers. Trop Anim Health Prod 51:523–528

Odior AO, Oyawale FA (2011) Application of time study model in rice milling firm: a case study. J Appl Sci Environ Manag 15:501–505

Ohtsubo K, Suzuki K, Yasui Y et al (2005) Bio-functional components in the processed pre-germinated brown rice by a twin-screw extruder. J Food Compos Anal 18:303–316

Omary MB, Fong C, Rothschild J et al (2012) Effects of germination on the nutritional profile of gluten-free cereals and pseudocereals: a review. Cereal Chem 89:1–14

Omidvar M, Karimi K, Mohammadi M (2016) Enhanced ethanol and glucosamine production from rice husk by NAOH pretreatment and fermentation by fungus Mucor hiemalis. Biofuel Res J 3:475–481

Onuma K, Kanda Y, Suzuki Ikeda S et al (2015) Fermented brown rice and rice bran with *aspergillus oryzae* (FBRA) prevents inflammation-related carcinogenesis in mice, through inhibition of inflammatory cell infiltration. Nutrients 7:10237–10250

Onyeneho S, Hettiarachehy N (1992) Antioxidant activity of durum wheat bran. J Agric Food Chem 40:1496–1500

Organization for Economic Co-Operation and Development (2004) Consensus document on compositional considerations for new varieties of rice (Oryza sativa): key food and feed nutrients and anti-nutrients. In: Annual report 2004, Organization for Economic Co-Operation and Development, Paris, France

Panche AN, Diwan AD, Chandra SR (2016) Flavonoids: an overview. J Nutr Sci 5:e47

Pandey R, Shrivastava SL (2018) Comparative evaluation of rice bran oil obtained with two-step microwave assisted extraction and conventional solvent extraction. J Food Eng 218:106–114

Pandey S, Byerlee D, Dawe D et al (eds) (2010) Rice in the global economy: strategic research and policy issues for food security. Los Baños, International Rice Research Institute

Papageorgiou M, Skendi A (2018) Introduction to cereal processing and by-products, sustainable recovery and reutilization of cereal processing by-products. Woodhead Publishing, Cambridge

Patel RM (2007) Stabilized rice bran. The functional food of the 21st century. Agro Food Ind Hi Tech 18:39–41

Paucar-Menacho LM, Martínez-Villaluenga C, Dueñas M et al (2017) Optimization of germination time and temperature to maximize the content of bioactive compounds and the antioxidant activity of purple corn (*Zea mays* L.) by response surface methodology. LWT Food Sci Technol 76:236–244

Peanparkdee M, Iwamoto S (2019) Bioactive compounds from by-products of rice cultivation and rice processing: extraction and application in the food and pharmaceutical industries. Trends Food Sci Technol 86:109–117

Peanparkdee M, Patrawart J, Iwamoto S (2019) Effect of extraction conditions on phenolic content, anthocyanin content and antioxidant activity of bran extracts from Thai rice cultivars. J Cereal Sci 86:86–91

Pham C-T, Boongla Y, Nghiem T-D et al (2019) Emission characteristics of polycyclic aromatic hydrocarbons and nitro-polycyclic aromatic hydrocarbons from open burning of rice straw in the north of Vietnam. Int J Environ Res Public Health 16:2343

Prabhakaran P, Ranganathan R, Muthu Kumar V et al (2017) Review on parameters influencing the rice breakage and rubber roll wear in sheller. Arch Metall Mater 62:1875–1880

Pradeep SR, Guha M (2011) Effect of processing methods on the nutraceutical and antioxidant properties of little millet (*Panicum sumatrense*) extracts. Food Chem 126:1643–1647

Pradhan A, Ali SM, Dash R (2013) Biomass gasification by the use of rice husk gasifier. Int J Adv Comput Theory Eng 2:14–17

Pratiwi R, Purwestri YA (2017) Black rice as a functional food in Indonesia. Funct Foods Health Dis 7:182–194

Quispe I, Navia R, Kahhat R (2017) Energy potential from rice husk through direct combustion and fast pyrolysis: a review. Waste Manag 59:200–210

Qureshi AA, Huanbiao M, Packer L et al (2000) Isolation and identification of novel tocotrienols from rice bran with hypocholesterolemic, antioxidant, and antitumor properties. J Agric Food Chem 48:3130–3140

Rafe A, Sadeghian A, Hoseini-Yazdi SZ (2017) Physicochemical, functional, and nutritional characteristics of stabilized rice bran form tarom cultivar. Food Sci Nutr 5:407–414

Rathod RP, Annapure US (2017) Physicochemical properties, protein and starch digestibility of lentil based noodle prepared by using extrusion processing. LWT Food Sci Technol 80:121–130

Ravichanthiran K, Ma ZF, Zhang H et al (2018) Phytochemical profile of brown rice and its nutrigenomic implications. Antioxidants (Basel) 7:71

Ray M, Ghosh K, Singh S et al (2016) Folk to functional: an explorative overview of rice-based fermented foods and beverages in India. J Ethnic Foods 3:5–18

Reddy VS, Dash S, Reddy AR (1995) Anthocyanin pathway in rice (*Oryza sativa* L.): identification of a mutant showing dominant inhibition of anthocyanins in leaf and accumulation of proanthocyanidins in pericarp. Theor Appl Genet 91:301–312

Rohman A, Helmiyati S, Hapsari M et al (2014) Rice in health and nutrition. Int Food Res J 21:13–24

Romani VP, Prentice C, Martins VG (2017) Active and sustainable materials from rice starch, fish protein and oregano essential oil for food packaging. Ind Crop Prod 97:268–274

Romasanta RR, Sander BO, Gaihre YK et al (2017) How does burning of rice straw affect CH_4 and N_2O emissions? A comparative experiment of different on-field straw management practices. Agric Ecosyst Environ 239:143–153

Roschat W, Theeranun S, Yoosuk B et al (2016) Rice husk-derived sodium silicate as a highly efficient and low-cost basic heterogeneous catalyst for biodiesel production. Energy Convers Manag 119:453–462

Saidelles APF, Senna AJT, Kirchner R et al (2012) Gestão de resíduos sólidos na indústria de beneficiamento de arroz [Solid waste management in rice processing industries]. Revista Eletrônica em Gestão, Educação e Tecnologia Ambiental 5:904–916

Samsudin NIP, Abdullah N (2014) Prevalence of viable *Monascus* van Tieghem species in fermented red rice (Hong Qu Mi) at consumer level in Selangor, Malaysia. J Biochem Microbiol Biotechnol 2:57–60

Serna-Saldivar SO (2016) 2-Physical properties, grading, and specialty grains, cereal grains, properties, processing, and nutritional attributes. CRC Press, Boca Raton

Sharif MK, Butt MS, Anjum FM et al (2014) Rice bran: a novel functional ingredient. Crit Rev Food Sci Nutr 54:807–816

Shin R, Jez JM, Basra A et al (2011) 14-3-3 proteins fine-tune plant nutrient metabolism. FEBS Lett 585:143–147
Shin HW, Jang ES, Moon BS et al (2016) Anti-obesity effects of gochujang products prepared using rice koji and soybean meju in rats. J Food Sci Technol 53:1004–1013
Siebenmorgen TJ, Nehus ZT, Archer TR (1998) Milled rice breakage due to environmental conditions. Cereal Chem 75:149–152
Singh Gujral H, Singh N (2002) Extrusion behaviour and product characteristics of Brown and milled rice grits. Int J Food Prop 5:307–316
Singh D, Kumar R, Kumar A et al (2008) Synthesis and characterization of rice husk silica, silica-carbon composite and H3PO4 activated silica. Cerâmica 54:203–212
Singh A, Rehal J, Kaur A et al (2015) Enhancement of attributes of cereals by germination and fermentation: a review. Crit Rev Food Sci Nutr 55:1575–1589
Singh R, Srivastava M, Shukla A (2016) Environmental sustainability of bioethanol production from rice straw in India: a review. Renew Sust Energy Rev 54:202–216
Smuda SS, Mohsen SM, Olsen K et al (2018) Bioactive compounds and antioxidant activities of some cereal milling by-products. J Food Sci Technol 55:1134–1142
So WKW, Law BMH, Law PTW et al (2018) A pilot study to compare two types of heat-stabilized rice bran in modifying compositions of intestinal microbiota among healthy Chinese adults. Adv Mod Oncol Res 4:253
Srichamnong W, Thiyajai P, Charoenkiatkul S (2016) Conventional steaming retains tocols and γ-oryzanol better than boiling and frying in the jasmine rice variety Khao dok mali 105. Food Chem 191:113–119
Steinkraus KH (1998) Bio-enrichment: production of vitamins in fermented foods. In: Wood JB (ed) Microbiology of fermented foods. Blackie Academic and Professional, London, pp 603–619
Surarit W, Jansom C, Lerdvuthisopon N et al (2015) Evaluation of antioxidant activities and phenolic subtype contents of ethanolic bran extracts of Thai pigmented rice varieties through chemical and cellular assays. Int J Food Sci Technol 50:990–998
Suryana R, Iriani Y, Nurosyid F et al (2018) Characteristics of silica rice husk ash from Mojogedang Karanganyar Indonesia. IOP Conf. Series. Mater Sci Eng 367:012008
Tamang JP (2015) Health benefits of fermented foods and beverages. CRC Press, New York
Tan BL, Norhaizan ME (2017) Scientific evidence of rice by-products for cancer prevention: chemopreventive properties of waste products from rice milling on carcinogenesis *in vitro* and *in vivo*. Biomed Res Int 2017:9017902. 18p
Tanaka K, Ogawa M, Kasai Z (1977) The rice scutellum. II. A comparison of scutellar and aleurone electron-dense particles by transmission electron microscopy including energy dispersive X-ray analysis. Cereal Chem 54:684–689
Tateoka T (1963) Taxonomic studies of Oryza. III. Key to the species and their enumeration. Bot Mag Tokyo 76:165–173
Thakur S, Singh PK, Das A et al (2015) Extensive sequence variation in rice blast resistance gene Pi54 makes it broad spectrum in nature. Front Plant Sci 6:345
The Rice Department MoAaC (2018) Thailand rice cultivation areas. www.ricethailand.go.th/rkb3/Eb_024.pdf
Ti H, Zhang R, Zhang M et al (2015) Effect of extrusion on phytochemical profiles in milled fractions of black rice. Food Chem 178:186–194
Tian S, Sun Y, Chen Z et al (2019) Functional properties of polyphenols in grains and effects of physicochemical processing on polyphenols. J Food Qual 2019:2793973. 8p
Tiwari BK, O'Donnell CP, Muthukumarappan K et al (2009) Anthocyanin and color degradation in ozone treated blackberry juice. Innov Food Sci Emerg Technol 10:70–75
Tong KT, Vinai R, Soutsos MN (2018) Use of Vietnamese rice husk ash for the production of sodium silicate as the activator for alkali-activated binders. J Clean Prod 201:272–286
Topakas E, Vafiadi C, Christakopoulos P (2007) Microbial production, characterization and applications of feruloyl esterases. Process Biochem 42:497–509

Torbica A, Škrobot D, Hajnal EJ et al (2019) Sensory and physico-chemical properties of whole grain wheat bread prepared with selected food by-products. LWT 114:108414

Towo EE, Svanberg U, Ndossi GD (2003) Effect of grain pretreatment on different extractable phenolic groups in cereals and legumes commonly consumed in Tanzania. J Sci Food Agric 83:980–986

Toyokuni S, Itani T, Morimitsu Y et al (2002) Protective effect of coloured rice over white rice on Fenton reaction-based renal lipid peroxidation in rats. Free Radic Res 35:583–592

Tylewicz U, Nowacka M, Martín-García B et al (2018) Target sources of polyphenols in different food products and their processing by-product: polyphenols properties, recovery, and applications. Woodhead Publishing, Cambridge

United States Department of Agriculture (2009) United States standards for rice. https://www.gipsa.usda.gov/fgis/standards/ricestandards.pdf. Accessed 19 Feb 2019

United States Department of Agriculture (2016) Overview of rice. https://www.ers.usda.gov/topics/crops/rice/. Accessed 19 Feb 2019

Unrean P, Fui BCL, Rianawati E et al (2018) Comparative techno-economic assessment and environmental impacts of rice husk-to-fuel conversion technologies. Energy 151:581–593

Vaisali C, Charanyaa S, Belur PD, Regupathi I (2015) Refining of edible oils: a critical appraisal of current and potential technologies. Int J Food Sci Technol 50:13–23

Van Dalen G (2004) Determination of the size distribution and percentage of broken kernels of rice using flatbed scanning and image analysis. Food Res Int 37:51–58

Vayghan AG, Khaloo AR, Rajabipour F (2013) The effects of a hydrochloric acid pre-treatment on the physicochemical properties and pozzolanic performance of rice husk ash. Cem Concr Comp 39:131–140

Wang T, Xue Y, Zhou M et al (2017) Comparative study on the mobility and speciation of heavy metals in ashes from co-combustion of sewage sludge/dredged sludge and rice husk. Chemosphere 169:162–170

Wang T, Khir R, Pan Z et al (2017a) Simultaneous rough rice drying and rice bran stabilization using infrared radiation heating. LWT 78:281–288

Witono JR, Juliani J (2019) Improving the resistance starch of rice through physical and enzymatic process. In: The 25th regional symposium on chemical engineering (RSCE 2018), vol. 268, p 01007, 4p

Wolfe KL, Liu RH (2003) Apple peels as a value-added food ingredient. J Agric Food Chem 51:1676–1683

Xia M, Ling WH, Ma J et al (2003) Supplementation of diets with the black rice pigment fraction attenuates atherosclerotic plaque formation in apolipoprotein E deficient mice. J Nutr 133:744–751

Xiong L, Sekiya EH, Sujaridworakun P et al (2009) Burning temperature dependence of rice husk ashes in structure and property. J Met Mater Miner 19:95–99

Xu E, Pan X, Wu Z et al (2016) Response surface methodology for evaluation and optimization of process parameter and antioxidant capacity of rice flour modified by enzymatic extrusion. Food Chem 212:146–154

Yağcı S, Göğüş F (2008) Response surface methodology for evaluation of physical and functional properties of extruded snack foods developed from food-by-products. J Food Eng 86:122–132

Yang I, S-h K, Sagong M et al (2016) Fuel characteristics of agropellets fabricated with rice straw and husk. Korean J Chem Eng 33:851–857

Yanık DK (2017) Alternative to traditional olive pomace oil extraction systems: microwave-assisted solvent extraction of oil from wet olive pomace. LWT 77:45–51

Yi B, Kim M-J (2019) Extraction of γ-oryzanol from rice bran using diverse edible oils: enhancement in oxidative stability of oils. Food Sci Biotech 29:393–399

Zain MFM, Islam MN, Mahmud F et al (2011) Production of rice husk ash for use in concrete as a supplementary cementitious material. Constr Build Mater 25:798–805

Zhai FH, Wang Q, Han JR (2015) Nutritional components and antioxidant properties of seven kinds of cereals fermented by the basidiomycete Agaricus blazei. J Cereal Sci 65:202–208

Zhao G, Zhang R, Dong L et al (2018) Particle size of insoluble dietary fiber from rice bran affects its phenolic profile, bioaccessibility and functional properties. LWT 87:450–456
Zheng B, Wang H, Shang W et al (2018) Understanding the digestibility and nutritional functions of rice starch subjected to heat-moisture treatment. J Funct Foods 45:165–172
Zhu H (2002) Utilization of rice bran by Pythium irregulare for lipid production. Master's thesis, Department of Biological and Agricultural Engineering, Agricultural and Mechanical College, Louisiana State University, USA
Zhu Y, Sang S (2017) Phytochemicals in whole grain wheat and their health-promoting effects. Mol Nutr Food Res 61:1600852

Chapter 4
Phytonutrients and Antioxidant Properties of Rice By-products

Abstract Rice by-products are good sources of phytochemicals and antioxidant activities. The richness and diversity in the bioactive compounds varied based on the extraction process and cultivars. Rice by-products are rich in tocopherols, tocotrienols, γ-oryzanol, phytic acid, phenolic compounds, γ-aminobutyric acid (GABA), and dietary fiber. These substances have been reported to have hypocholesterolemic, hypotensive, anticancer, and anti-diabetic activities. Most of these bioactive constituents show encouraging findings in the prevention of these diseases. Nonetheless, the bioactivities of these bioactive compounds are poorly understood. In this chapter, we presented the composition of bioactive components in rice by-products. The results from both *in vivo* and *in vitro* experiments in relation to chronic diseases are also discussed in this chapter.

Keywords Vitamin E · Gamma-oryzanol · γ-aminobutyric acid · Phytic acid · Antioxidant activity · Dietary fiber

Rice by-products have been reported to decrease the development of chronic diseases (Peanparkdee and Iwamoto 2019). Most of the studies suggest the preventive or therapeutic activities of rice by-products were due to the antioxidant activity and the ability to scavenge the free radicals. Therefore, tremendous effort has been made to evaluate the nutritional properties of rice by-products as a source of dietary fiber and other health-promoting phytochemical compounds. Besides bioactive compounds, rice by-products also contain various nutrients and mineral composition (Table 4.1).

4.1 Vitamin E

Vitamin E is a pivotal antioxidant found naturally in food. Vitamin E comprises of a group of eight structurally related lipophilic chromanol congeners, including four tocotrienols and four tocopherols (Fig. 4.1). Both tocotrienols and tocopherols are

B. L. Tan, M. E. Norhaizan, *Rice By-products: Phytochemicals and Food Products Application*, https://doi.org/10.1007/978-3-030-46153-9_4

Table 4.1 Proximate composition and mineral content of rice by-products

Nutrients	Broken rice[a]	Rice husk[b]	Rice bran[c]	Polishings[d]	Rice straw[e]
Dry matter	87.0–89.0	87.0–92.5	89–94	90.0	90.9
Protein[f]	6.7–9.8	2.1–4.3	10.6–16.9	11.2–13.4	1.2–7.5
Crude fat	0.5–1.9	0.30–0.93	5.1–19.7	10.1–13.9	0.8–2.1
Crude fiber	0.6	30.0–53.4	7.0–18.9	2.3–3.6	33.5–68.9
Ash	5.0	13.2–24.4	8.8–28.8	5.2–8.3	12.2–21.4
Carbohydrate	–	22.4–35.3	90.0	51.1–55.0	39.1–47.3
Calcium	0.09–0.19	0.04–0.21	0.08–1.4	0.05	0.30–0.71
Phosphorus	0.03–0.04	0.07–0.08	1.3–2.9	1.48	0.06–0.16

[a]Farrell and Hutton (1990); NGFA (National Grain and Feed Association) (2003); NRC (National Research Council) (1982); NRC (1994); NRC (1998)
[b]AgrEvo (1999); Farrell and Hutton (1990); Ffoulkes (1998); FAO (2003); Herd (2003); Juliano and Bechtel (1985); Miller et al. (1991); NGFA (National Grain and Feed Association) (2003)
[c]AgrEvo (1999); Farrell and Hutton (1990); Ffoulkes (1998); FAO (2003); Herd (2003); Juliano and Bechtel (1985); Miller et al. (1991); NGFA (National Grain and Feed Association) (2003); NRC (National Research Council) (1982); NRC (1994); NRC (1998); NRC (2000); NRC (2001)
[d]Miller et al. (1991); NRC (1994); NRC (1998)
[e]Drake et al. (2002); Fadel and MacKill (2002); FAO (2003); Ffoulkes (1998); Wanapat et al. (1996); Nour (2003)
[f]Animal scientists commonly use a conversion factor of N × 6.25 for crude protein (AOAC (Association of Official Analytical Chemists) 2002)

further divided into α-, β-, γ-, and δ- according to the hydroxyl and methyl substitution in their phenolic ring (Bakir et al. 2020). Alpha-tocopherol is a fat-soluble compound, in which antioxidative capacity has been widely studied. It exerts numerous biological activities including prevents lipid peroxidation and propagation and deactivates photosynthesis-derived reactive oxygen species (ROS), particularly $^{\bullet}O_2^-$ through scavenging of lipid peroxyl radicals (Nayak et al. 2019). Alpha-tocopherol, the most active form of vitamin E found predominantly in rice germ; while γ-tocopherol is primarily found in rice bran (Esa et al. 2013). Intriguingly, the vitamin E level in rice germ is 5 times higher than rice bran. In support of this, Moongngarm et al. (2012) found that rice germ contains high levels of vitamin E, particularly γ- and α-tocopherols, compared to rice bran. The previous study found that infrared radiation heating (140 °C for 15 min) can significantly reduce the vitamin E of stabilized rice bran (Irakli et al. 2018), suggesting the importance of optimum infrared radiation condition to improve the quality of vitamin E content. The tocopherol levels in methanol extract of defatted rice bran from India (138 μg/g) (Renuka Devi and Arumughan 2007) are lower than that of Thailand cultivars (350–670 μg/g) (Chotimarkorn et al. 2008). In a study by Lai et al. (2009) evaluating tocopherols and tocotrienols on methanol, ethyl acetate, and hexane extracts of *Japonica* rice bran, ethyl acetate extract of *Japonica* rice bran showed the highest vitamin E levels (770 μg/g) compared to methanol (573 μg/g) and hexane (196 μg/g) extracts. The amount of tocopherols (27.4–129.6 μg/g) and tocotrienols (20.8–301.7 μg/g) in 80% methanol extract of whole rice bran was much lower than that of absolute methanol extract (Forster et al. 2013). Comparing with rice bran, the

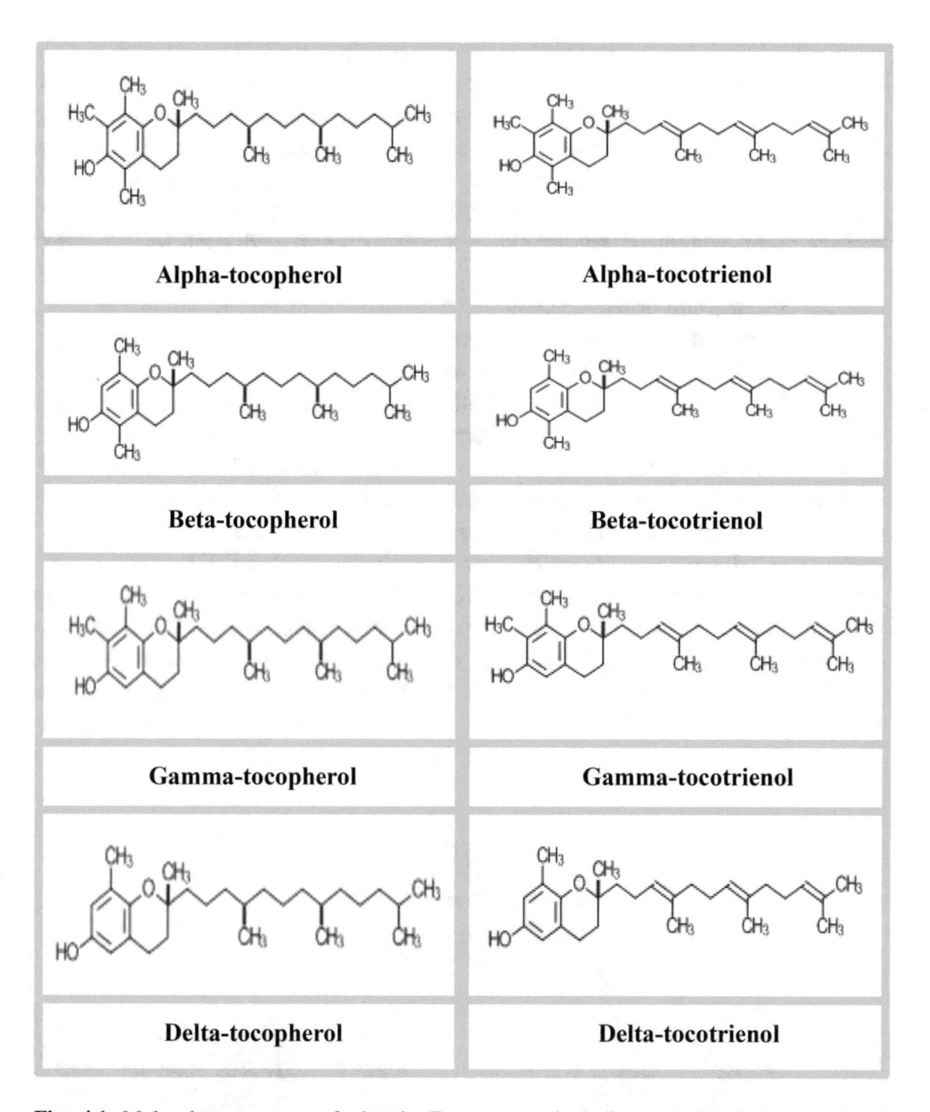

Fig. 4.1 Molecular structures of vitamin E congeners including tocopherols (α-tocopherol, β-tocopherol, γ-tocopherol, and δ-tocopherol) and tocotrienols (α-tocotrienol, β-tocotrienol, γ-tocotrienol, and δ-tocotrienol)

amount of tocopherols (3.4 μg/g) and tocotrienols (3.25 μg/g) in brewers' rice was lower than that found in rice bran (Tan et al. 2013).

Several protein kinases, especially protein kinase C (PKC) sub-family members can be mediated in human neuronal cells administered with tocotrienols and tocopherols (Galli et al. 2017). The signaling affects the cell cycle regulation and apoptotic cell death in different cell lines, including human glioblastoma and neuron cells with a strong γ-tocopherol and α-tocopherol on the phosphorylative activation of pro-survival mitogen-activated protein kinase/extracellular regulated protein kinases (MAPK-ERK) isoforms. α-tocopherol also protects mouse cortical and hippocampal neurons from apoptosis through regulation of neurodegenerative signaling cascades and thus preserves the function of brain (Ambrogini et al. 2014). In addition, α-tocotrienol also involved in the modulation of 12-lipoxygenase and phospholipase A2 activities (Sen et al. 2007; Khanna et al. 2010). Figure 4.2 shows the effects of vitamin E on neurodegenerative disease and cardiovascular disease (CVD).

A specific therapy or higher administration of vitamin E is vitally important in the developing brain in relation to inflammatory activation or damage of glial components. The hippocampus has been demonstrated to be an elective area for the physiological and even therapeutic potentials in the developing brain. In addition to the effects observed in the developing brain, vitamin E has also been suggested in mediation of other phases in brain with anti-inflammatory properties (Naito et al. 2005; Reiter et al. 2007). The circulating levels and the intake of different forms of vitamin E have been reported to influence Alzheimer's disease and mild cognitive impairment (Mangialasche et al. 2013a; Mangialasche et al. 2013b). Amyloid-beta peptides can oxidize proteins with specific amino acid residues and subsequently lead to an elevation of oxidative stress in Alzheimer's disease brains (Butterfield and Boyd-Kimball 2005). Administration of vitamin E in rat neuronal cultures inhibits the formation of amyloid beta-associated ROS and reduces the oxidative stress markers (Yatin et al. 2000). Table 4.2 summarized the antioxidant and phytochemicals properties of rice by-products in animal and human studies.

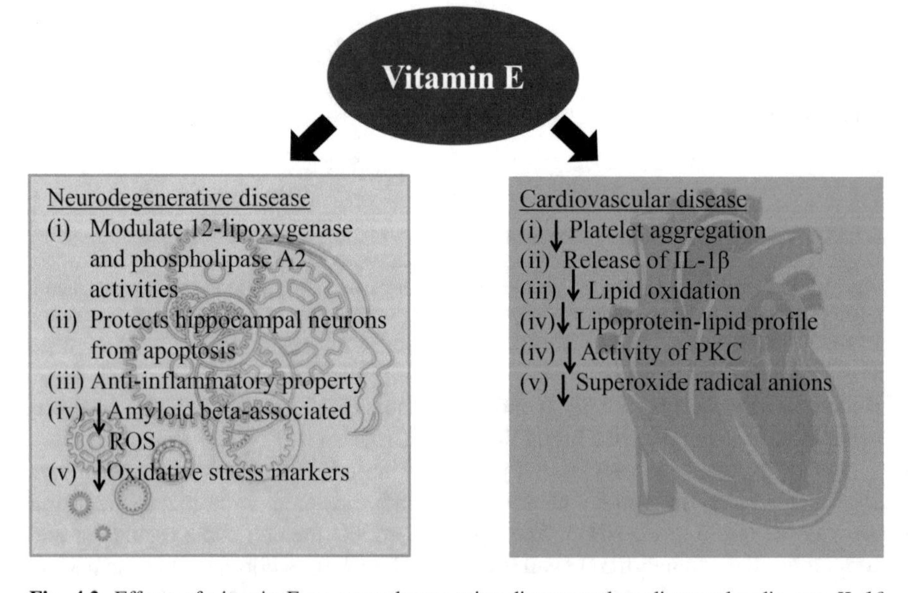

Fig. 4.2 Effects of vitamin E on neurodegenerative disease and cardiovascular disease. *IL-1β* interleukin-1beta, *PKC* protein kinase C, *ROS* reactive oxygen species

Table 4.2 Phytochemicals and antioxidant properties of rice by-products

Phytochemical	Diseases/metabolic disorders	Findings	References
Tocotrienols and tocopherols	Neurodegenerative disease	↓ Formation of amyloid beta-associated ROS	Yatin et al. (2000)
		Protect mouse cortical and hippocampal neurons from apoptosis	Ambrogini et al. (2014)
	Cardiovascular disease	↓ Formation of cholesterol-induced atherosclerotic lesions	Özer et al. (1998)
		↓ Platelet aggregation and lipid oxidation	Devaraj and Jialal (1998)
		↓ Lipoprotein-lipid profile	Budin et al. (2009)
Gamma-oryzanol	Cardiovascular disease	↓ Platelet aggregation and total plasma cholesterol	Cicero and Gaddi (2001)
		↑ Expression of IL-10 and IL-4 compared to control group	Rao et al. (2016b)
		↓ Secretion of IL-1β by peritoneal macrophages	Rao et al. (2016a)
	Cancer	↓ Inflammatory response in mice-induced colitis	Islam et al. (2008)
		↓ Inflammatory response by reducing NF-κB transcriptional activity	Islam et al. (2011)
	Type 2 diabetes mellitus	Improved blood adiponectin concentrations via activation of PPARγ	Islam et al. (2009)
γ-aminobutyric acid	Cardiovascular disease	↓ Risk of cardiovascular disease	Roohinejad et al. (2010)
		↓ The elevation of cholesterol in human	Inoue et al. (2003)
	Type 1 diabetes	↓ Risk of type 1 diabetes	Ligon et al. (2007)
	Type 2 diabetes mellitus	↑ Insulin secretion from the pancreas	Adeghate and Ponery (2002)
		Improved hyperglycemia	Chen et al. (2016)
	Hypertension	↓ Blood pressure in human	Inoue et al. (2003)
		↓ Blood pressure in hypertensive rats	Yamakoshi et al. (2007); Tsai et al. (2013)
		Only observed during acute administration but no effect after chronic administration	Yang et al. (2012)
	Depression	↓ Levels of GABA in the occipital cortex and dorsolateral prefrontal in depressed patients	Hasler et al. (2007); Bhagwagar et al. (2007); Price et al. (2009)

(continued)

Table 4.2 (continued)

Phytochemical	Diseases/metabolic disorders	Findings	References
	Anxiety	↓ Anxiety and induces relaxation	Abdou et al. (2006)
Phytic acid	Cancer	↓ Incidence of colonic cancer	Graf and Eaton (1990)
	Renal stone	↓ Crystallization of calcium oxalate salts in the urine	Ahmed et al. (2016)
	Type 2 diabetes mellitus	Improved insulin release	Dou et al. (2014)
Polyphenols	Cardiovascular disease	Prevent cardiovascular disease	Nunes et al. (2016)
		Exert pro-oxidant effect in Cu^{2+}-induced oxidation of LDL	Barnaba and Medina-Meza (2019)
		↓ Nearly 6–7% deaths from CVD	Lai et al. (2015)
		↓ Lipid peroxidation, oxidative stress, and total TG levels in macrophages in the aortas	Rosenblat et al. (2015)
	Cancer	Induced apoptotic cell death	Ryan et al. (2011)
	Type 2 diabetes mellitus	↓ Blood glucose in diet-induced diabetic ICR mice and S961-treated C57BL/6 mice	Huang et al. (2019)
		Improved insulin action and β-cell function	Jung et al. (2004)
		↓ α-glycosidase, α-amylase, and AR enzymes	Demir et al. (2019)

AR aldose reductase, *CVD* cardiovascular disease, *GABA* γ-aminobutyric acid, *ICR* imprinting control region, *IL-1β* interleukin-1beta, *IL-4* interleukin-4, *IL-10* interleukin-10, *LDL* low-density lipoprotein, *NF-κB* nuclear factor-kappa B, *PPARγ* peroxisome proliferator-activated receptor-γ, *ROS* reactive oxygen species, *TG* triglycerides

Substantial evidence highlights the beneficial role of vitamin E against lipid peroxidation of cell membranes and its ability in inhibition of radical chain by forming a low-reactivity derivative (Galli et al. 2017) and subsequently prevents cell membranes against free radical damage activated by low-density lipoprotein (LDL). Vitamin E can positively alter the oxidative stress biomarkers, enhance erythropoiesis, and reduce erythropoietin concentration (Niki 2014). Research evidence has shown that a high intake of vitamin E was inversely associated with proatherogenic. These favorable effects are mediated by the release of interleukin-1beta (IL-1β) and superoxide radical anions via modulation of platelet aggregation, lipid oxidation, and smooth muscle cell proliferation (Devaraj and Jialal 1998).

RRR-α-tocopherol inhibits the proliferation of LDL stimulated smooth muscle cells by modulating PKC activity *in vitro* (Özer et al. 1993). Feeding rabbits with cholesterol diets increased the activity of PKC and atherosclerotic lesions in the aortic tissues. By contrast, vitamin E prevents the induction of PKC activity and the formation of cholesterol-induced atherosclerotic lesions, suggesting that several signaling transduction pathways took part in the protective event of vitamin E against atherosclerosis (Özer et al. 1998). An animal study has revealed that tocotrienol supplementation decreased lipoprotein-lipid profile, with a reduction in key markers of lipid and protein damage (Budin et al. 2009). The improvements in these indices could be due to the alteration of gene expression through several signaling pathways including β-hydroxy-β-methylglutaryl coenzyme A reductase activity (Holifa et al. 2017). Further, tocotrienol supplementation can also decrease DNA damage in healthy elderly (Ng and Ko 2012).

4.2 Gamma-oryzanol

Gamma-oryzanol is a mixture of triterpene alcohols and ferulic acid esters of sterol (Patel and Naik 2004), which exerts potent antioxidant activity. Among the phytosteryl ferulate esters, campesteryl ferulate, 24-methylenecycloartanyl ferulate, and cycloartenyl ferulate have been identified as the predominant components of γ-oryzanol (Xu and Godber 1999) (Fig. 4.3). The previous study stated that γ-oryzanol of Thai glutinous purple rice bran from different cultivars was in the range of 1.23–9.14% (Saenjum et al. 2012). The study further demonstrated that Thai glutinous purple rice bran is an efficient scavenger of superoxide anion radical (Saenjum et al. 2012). However, Juliano et al. (2005) found that γ-oryzanol did not scavenge superoxide anion radical. Interestingly, rice germ has 5 times lower in γ-oryzanol compared to rice bran (Yu et al. 2007). Consistent with the study obtained by Yu et al. (2007), Butsat and Siriamornpun (2010) also found that bran contains the highest level of oryzanol compared to other parts of rice. Aqueous riceberry broken rice extract (52.5 mg/100 g) has lower γ-oryzanol content compared to ethanolic crude riceberry rice oil (234 mg/100 g). Among all the isomers, cycloartanyl

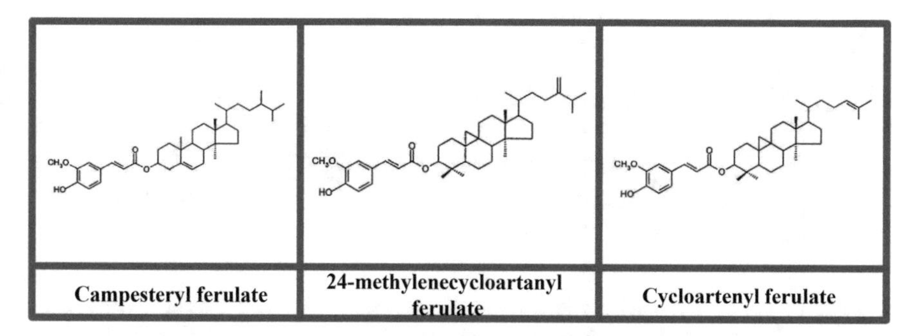

Fig. 4.3 Molecular structures of predominant components of γ-oryzanol including campesteryl ferulate, 24-methylenecycloartanyl ferulate, and cycloartenyl ferulate

ferulate (56.3%) is one of the major γ-oryzanol isomers present in riceberry broken rice extract, followed by campesteryl ferulate (27.5%), 24-methylene cycloartanyl ferulate (16.2%), and β-sitosteryl ferulate (below limit of detection) (Luang-In et al. 2018). Moongngarm et al. (2012) studied γ-oryzanol of rice germ, rice bran, and rice bran layer in four indica rice cultivars (non waxy and waxy). The data showed that γ-oryzanols in rice germ (1.41–1.75 mg/g) are lower than that of the rice bran (3.5–9.12 mg/g) (Moongngarm et al. 2012).

Gamma-oryzanol has been reported to possess health benefits including reducing platelet aggregation, increasing high-density lipoprotein cholesterol (HDL-C), decreasing total plasma cholesterol, and improving plasma lipid profile (Cicero and Gaddi 2001) (Fig. 4.4). Furthermore, substantial evidence also highlights that γ-oryzanol displayed antioxidant properties (Kartikawati and Purnomo 2019). The macrophage from rats fed a diet containing RBO reduced the expression of proinflammatory cytokines such as tumor necrosis factor-alpha (TNF-α) and interleukin-6 (IL-6) and stimulated anti-inflammatory cytokines, for instance, IL-4 and IL-10 (Lee et al. 2019). Notably, feeding rats containing RBO + minor constituents removed + γ-oryzanol secreted higher expression of IL-10 and IL-4 compared to groundnut oil group (control), suggesting that γ-oryzanol may contribute to the anti-inflammatory activity of RBO (Rao et al. 2016b). It has been demonstrated that

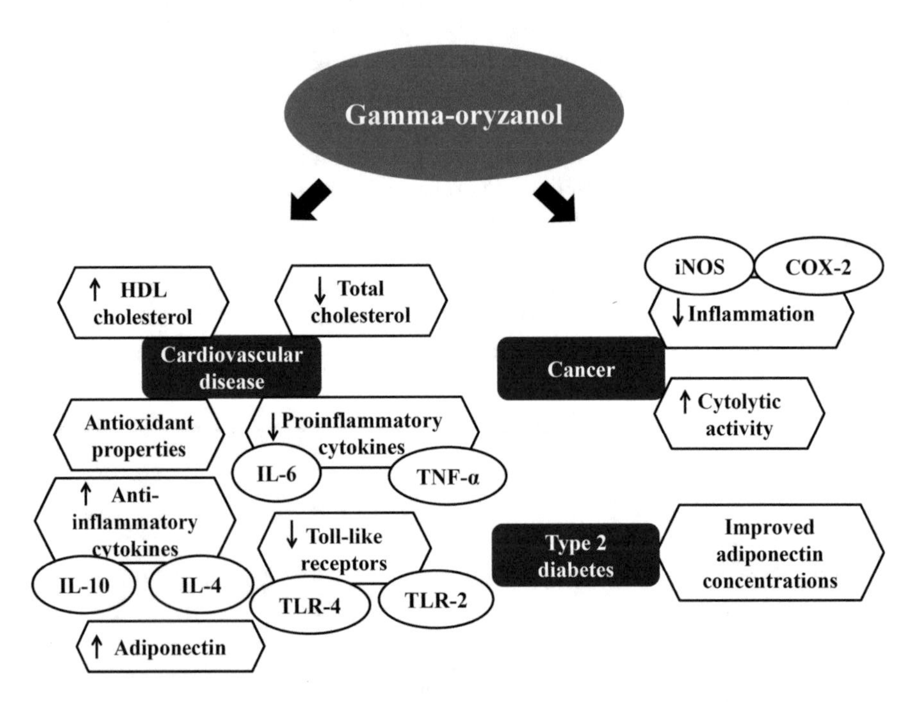

Fig. 4.4 Effect of γ-oryzanol on cardiovascular disease, cancer, and type 2 diabetes. *COX-2* cyclooxygenase-2, *HDL* high-density lipoprotein cholesterol, *IL-4* interleukin-4, *IL-6* interleukin-6, *IL-10* interleukin-10, *iNOS* inducible nitric oxide synthase, *TLR-2* toll-like receptor-2, *TLR-4* toll-like receptor-4, *TNF-α* tumor necrosis factor-alpha

stimulation of nuclear factor-kappa B (NF-κB) induces the proinflammatory signaling pathway and subsequently resulting in the overproduction of matrix metalloproteases, cytokines, eicosanoids, lysosomal enzyme, and ROS by macrophages (Sakai et al. 2012). An animal study showed that RBO containing γ-oryzanol significantly reduced the secretion of IL-1β by peritoneal macrophages compared to rats fed with hydrogenated fat (Rao et al. 2016a). RBO also reduced the expression of NF-κBp65 and toll-like receptors (TLR-4 and TLR-2) and increased the expression of adiponectin in macrophages (Rao et al. 2016a), indicated that the anti-inflammatory properties of RBO may be modulated through downregulating of NF-κBp65 expression which reduces the proinflammatory mediators.

In addition, cycloartenyl trans-ferulate and γ-oryzanol markedly inhibited the inflammatory response in mice-induced colitis (Islam et al. 2008). Islam et al. (2011) further showed that rice bran phytosteryl ferulates inhibit inflammatory response by reducing NF-κB transcriptional activity and thereby downregulating inducible nitric oxide synthase (iNOS) and cyclooxygenase-2 (COX-2) inflammatory enzymes and proinflammatory cytokines (Fig. 4.4). In addition, rice bran phytosteryl ferulates also improved blood adiponectin concentrations through stimulation of peroxisome proliferator-activated receptor-γ (PPARγ) and inhibition of NF-κB (Islam et al. 2009) (Fig. 4.4). Overall, rice bran γ-oryzanol suppressed tumor proliferation by phagocytosis in peritoneal macrophages, partial restoration of NO production, and activating the cytolytic activity in splenic natural killer (NK) cells, and thus leading to the release of proinflammatory cytokines from macrophages such as TNF-α, IL-1β, and IL-6.

4.3 γ-aminobutyric Acid

Gamma-aminobutyric acid (GABA) is a four-carbon non-protein amino acid existing in nature (Fig. 4.5). It is synthesized mainly through the α-decarboxylation of glutamic acid catalyzed by glutamate decarboxylase (GAD) (Yin et al. 2018). It has been demonstrated that GABA is ubiquitous among plants and its amounts in plant

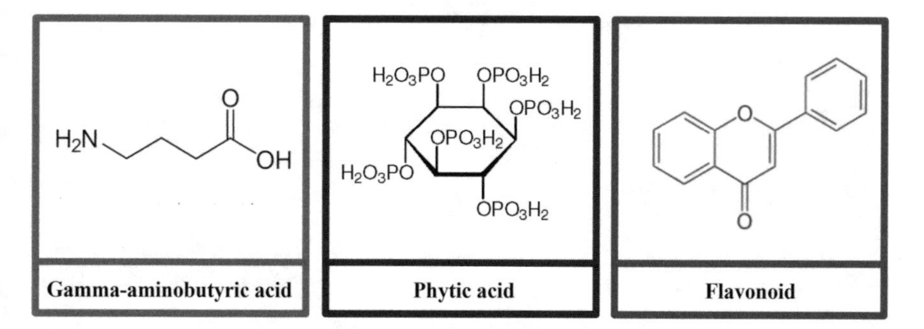

Fig. 4.5 Molecular structures of γ-aminobutyric acid, phytic acid, and flavonoid

tissues are elevated in response to stress during the processing of seeds including germination and soaking or during the processing of seeds (Xu and Hu 2014; Diana et al. 2014). Some other processes have been found to enhance the concentration of GABA in plant materials such as fermentation, pre-germination, gaseous treatment, and enzymatic treatment (Roohinejad et al. 2011). Substantial amounts of neurotransmitter GABA are found in rice germ (Ahn et al. 2014). Cultivation of *Lactobacillus sakei* B2-16 in the rice bran extracts containing 12% monosodium glutamate, 1% yeast extract, and 4% sucrose increased 2.4-fold GABA levels (Kook et al. 2010).

GABA exerts several physiological roles including decreasing the risk of CVD via modulation of cholesterolemia (Roohinejad et al. 2010). An animal study has shown that GABA decreases the risk or reverses Type 1 diabetes (T cells autoimmunity) via regulation of pancreatic islet cells (Ligon et al. 2007), activation of the insulin secretion by modulating the positive autocrine feedback loop in human pancreatic β-cells through GABA-$GABA_A$ receptor system (Braun et al. 2010), and suppressing of inflammatory T cell response (Tian et al. 2004). In a study by Chen et al. (2016) evaluated the anti-diabetic effect of GABA-rich yogurt in relation to streptozotocin-induced diabetic mice. The data showed that yogurt enriched with GABA improved hyperglycemia and increased serum insulin levels. Similar dietary supplementation has also been found to enhance insulin secretion from the pancreas in the normal rats (Adeghate and Ponery 2002). In another study, Braun et al. (2004) found that GABA is involved in the release of glucagon. The study showed that the release of endogenous GABA from β-cells in rats suppresses the releases of insulin and glucagon through the stimulation of GABA receptors. Figure 4.6 shows the effects of GABA on types 1 and 2 diabetes, anxiety, and depression.

Dietary intake of food high in GABA has been reported to decrease the elevation of cholesterol and blood pressure in human (Inoue et al. 2003) and animals (Abe et al. 1995; Lacerda et al. 2003; Hayakawa et al. 2004). Feeding hypertensive rats

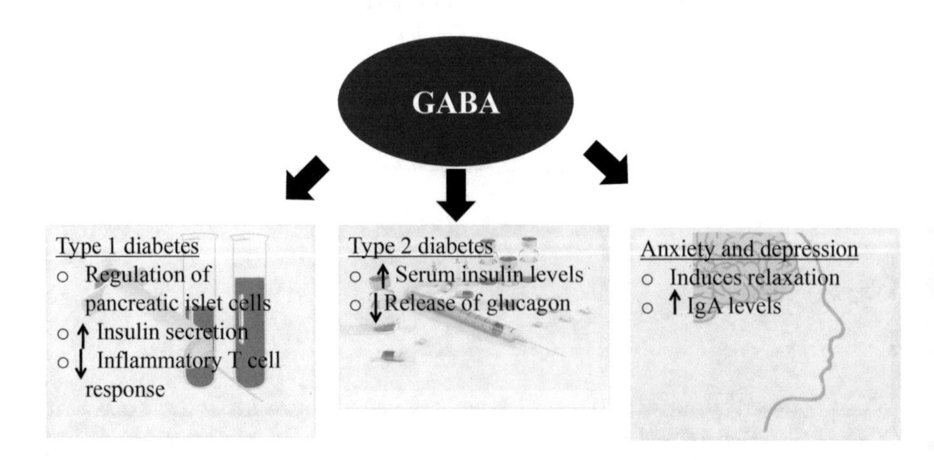

Fig. 4.6 Biological mechanism of γ-aminobutyric acid in relation to type 1 diabetes, type 2 diabetes, and anxiety and depression. *GABA* γ-aminobutyric acid, *IgA* immunoglobulin A

with reduced-sodium soy sauce rich in GABA for 6 weeks decreased blood pressure compared to a reduced-sodium soy sauce alone (Yamakoshi et al. 2007). Moreover, a similar finding has also been reported by Tsai et al. (2013), who found that hypertensive rats fed with GABA-rich Chingshey purple sweet potato fermented milk reduced blood pressure. These findings are further supported by Inoue et al. (2003) who revealed that dietary intake of fermented milk containing 10 mg GABA for 12 weeks decreases the blood pressure in hypertensive patients. However, Yang et al. (2012) reported that anti-hypertensive effect of GABA was only observed during acute administration, while none of the effects after chronic administration.

GABA has been reported in the modulation of tranquilizing, diuretic, and hypotensive activities, and thus may play a crucial role in inhibitory neurotransmitter in the central nervous system (Li and Cao 2010; Dhakal et al. 2012). Substantial data highlights the roles of GABA consumption in the control of anxiety and depression (Romeo et al. 2017). A study has shown a positive relationship between reduced levels of GABA in the occipital cortex and dorsolateral prefrontal in depressed patients (Hasler et al. 2007; Bhagwagar et al. 2007; Price et al. 2009). Indeed, GABA-A receptor plays an important role in regulating depression, phobias, fear, and anxiety (Gray et al. 1984; Coupland and Nutt 1995). GABA has been well-recognized as an effective compound to regulate several neurological disorders such as Parkinson's disease, Huntington's chorea, and Alzheimer's disease (Dargaei et al. 2018; Guzmán et al. 2018; Tuura et al. 2018). The data from the human study showed that GABA decreases anxiety, induces relaxation, and increases immunoglobulin A (IgA) levels in humans administered with GABA. This finding implied that GABA could work effectively by enhancing immunity as well as serve as a natural relaxant under stress conditions (Abdou et al. 2006). In this regard, the GABA production using rice bran is cost-effective to be utilized in the industrial fields (Kook et al. 2010).

4.4 Phytic Acid

Inositol hexaphosphate (IP6) also known as phytic acid is a bioactive compound (Fig. 4.5). It is the primary nutritionally related form of inositol existing in nature. Phytic acid is mainly found in the bran proportion of whole grain such as cereal, particularly located within the aleurone layer. Phytic acid in rice seeds is primarily used for the storage of phosphates and thereby served as a potent antioxidant and energy source (Silva and Bracarense 2016). Rice phytic acid is regarded as an anti-nutrient due to its propensity when interacting with minerals and thereby may result in a deficiency in humans that rely on these food sources as the key form of nutrition (Bohn et al. 2008). Phytic acid has long been considered to be nutritionally negative because it chelates minerals including Mg, Ca, Fe, and Zn, and thereby limiting the intestinal bioavailability (Lopez et al. 2002). It binds with free iron and thus inhibits iron-driven oxidative reactions (Wang et al. 2017).

Several studies have investigated the rice bran phytic acid using several techniques, for instance, colorimetric and high-performance liquid chromatography (HPLC) analyses, in which the phytic acid contents may vary based on different processing methods. For instance, the amounts of phytic acid in rice bran was accessed by titration method was 57.7 mg/g (Garcia-Estepa et al. 1999), HPLC and colorimetric (78 and 54 mg/g) (Knuckles et al. 1982), HPLC (60 mg/g) (Kasim and Edwards 1998), and colorimetric method (36.5 mg/g) (Ravindran et al. 1994). Compared to cereal, the phytic acid of Tarom cultivar rice bran was almost similar to oat bran (21.5–24.0 mg/g). By contrast, the content of phytic acid was less than that of some varieties of wheat bran (25–47 mg/g) (Garcia-Estepa et al. 1999). The amount of phytic acid (0.38 mg/g) in brewers' rice was lower than that of Tarom cultivar rice bran (Tan et al. 2013). In the context of phytic acid in rice germ (31.87–39.54 mg/g), the content of phytic acid in rice bran was higher than that of rice germ of different rice cultivars for both non-waxy and waxy rice, which ranged from 35.00 to 50.68 mg/g (Moongngarm et al. 2012). It seems like phytic acid levels of rice bran may be varied based on the state of bran separation and the place of cultivation. In addition, extrusion cooking can significantly reduce the phytates in rice bran, implied that this process could destroy the anti-nutritional matters in rice bran (Rafe et al. 2017).

Phytic acid has shown potent antioxidant *in vivo* and *in vitro* studies (Li et al. 2018; da Silva et al. 2019). It has demonstrated an ability to inhibit Fe catalyzed oxidative reactions due to its ability to chelate free Fe by decreasing lipid peroxidation (Graf et al. 1987). In the context of cancer, phytic acid decreases the incidence of colonic cancer by reducing oxidative damage on the gut epithelium, especially in the colon, in which the bacteria may produce oxygenated radicals (Graf and Eaton 1990) (Fig. 4.7). Furthermore, phytic acid also suppresses xanthine oxidases-induced superoxide-dependent DNA damage (Muraoka and Miura 2004). Xanthine oxidase produces superoxide anions (O_2^-) during the oxidation of xanthine in the intestine (Battelli et al. 1972). The previous study stated that phytic

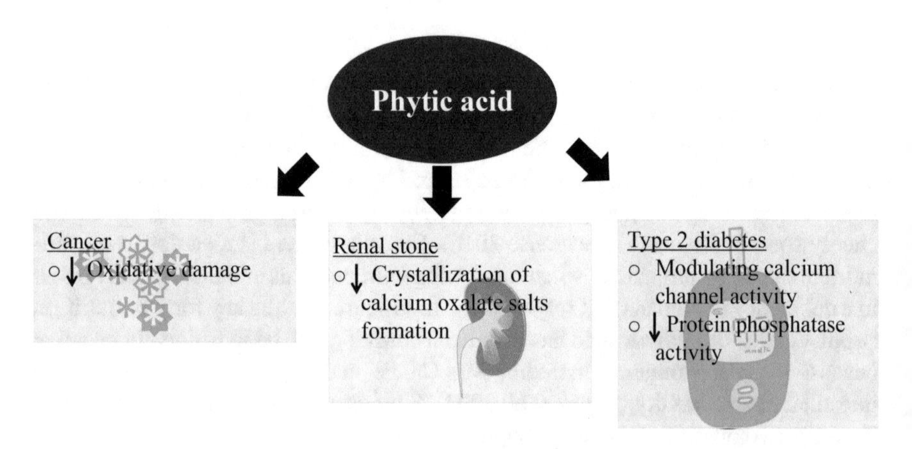

Fig. 4.7 Biological mechanism of phytic acid on cancer, renal stone, and type 2 diabetes

acid can suppress the crystallization of calcium oxalate salts in the urine and thereby could prevent the development of renal stone (Ahmed et al. 2016) (Fig. 4.7). Besides inhibiting the formation of renal stone and suppressing colon cancer, it has also been demonstrated as a crucial element in regulating insulin secretion, particularly through calcium channel activity (Fig. 4.7). Influx extracellular calcium is one of the events that improve insulin release (Dou et al. 2014). The molecular mechanisms of phytic acid that implicate the release of insulin are complex but it appears that it may suppress protein phosphatase activity and thus open intracellular calcium channels and ultimately enhance insulin release (Larsson et al. 1997; Barker and Berggren 1999).

4.5 Antioxidant Activity and Phenolic Compounds

There is some common extraction techniques that are employed for extracting bioactive constituents including the methods of solvent extraction, a conventional method used to extract bioactive constituents from rice. The polarity and type of solvents are critical for the solubility of bioactive compounds in rice. The extracting solvents that are commonly used are water, ethanol, methanol, and acetone (Imsanguan et al. 2008; Heinemann et al. 2008; Nam et al. 2006). Ethanol at a concentration of 50% was utilized as an extraction solvent at a portion of 1:5 bran or grain liquid, and the extraction was conducted for 3–12 h at room temperature. The extract was concentrated using a rotary evaporator until all ethanol residues were partitioned and removed with saturated butanol to obtain the medium polar bioactive constituents from the red rice extract (Limtrakul et al. 2016; Pintha et al. 2014; Pintha et al. 2015) or black rice extract (Limtrakul et al. 2015). In addition, Xia et al. (2006) optimized the extraction condition to obtain phytochemicals from black rice using 60% ethanol containing 0.1% hydrochloric acid as the extractant solvent at a 1:10 feed to liquid proportion. These techniques have low efficiency and may enhance the environmental pollution due to the large volume of organic solvents and the length of extraction time (Lee et al. 2013). A study by Renuka Devi and Arumughan (2007) found that using methanol at different bran-solvent ratios (1:5, 1:10, 1:15, and 1:20 (w/v) for 10 h) for extracting the phytochemicals from rice bran, the bran-solvent ratio of 1:20 (w/v) provided the highest yields of bioactive components, such as ferulic acid, oryzanol, and total phenolic contents. The data further showed that the extraction time also influences the extractability and the extraction yield of phytochemicals. Notably, about 97% of oryzanol was released at a faster rate after fourth hour. Zubair et al. (2012) evaluated the effective recovery of antioxidant activity and phenolics using different extraction solvents (pure and aqueous isopropanol, pure and aqueous ethanol, and pure and aqueous methanol) in brown rice. The results showed that aqueous methanol (20:80 water:methanol (v/v)) and aqueous isopropanol (20:80 water:isopropanol (v/v)) gave the higher ferrous ion-chelating activity, reducing power, total phenolic content, and DPPH free radical-scavenging capacity compared to other extraction solvents (Zubair et al. 2012). This finding implies that aqueous organic mixture of solvents, namely methanol and isopropanol are more effective in the recovery of rice antioxidant compounds such as metal chelators, free radical scavengers, and phenolics. Seesom et al. (2018) found that combination of solvent extraction, namely methanol/acetone/water (35:35:30 (v/v/v)), and pressurized liquid extraction gave the maximum yields of total phenolic acids in brown, Basmati (Basmati 370-1), and Jasmine (Jasmine 85) rice. The data further revealed that microwave alkaline hydrolysis is effective in extracting the bound phenolic acids (Seesom et al. 2018).

The previous study stated that water extract of riceberry broken rice exhibited DPPH radical scavenging activity (6.36 mg TEAC/g extract), total phenolic content (7.24 mg GAE/g extract), and total flavonoid content (25.7 mg CE/g extract) (Luang-In et al. 2018). The previous study stated that DPPH radical scavenging activity in rice germ of red rice and black rice are ranged from 1.03–1.15 mg/mL. While in brewers' rice (2.90–3.97 mg/mL) (Tan et al. 2013), the DPPH value is comparatively higher than the rice germ of red rice and black rice. Intriguingly, the EC_{50} value of several types of Thailand rice bran (0.38–0.74 mg/mL) (Chotimarkorn et al. 2008) was lower than those of the brewers' rice, broken rice, and rice germ, indicating that rice bran exerts greater ability in scavenging free radicals compared to other rice by-products. These data are in line with the earlier study obtained by Butsat and Siriamornpun (2010), who found that the husk and bran had higher antioxidant activity than brown or milled rice as evaluated by ferric reducing antioxidant power (FRAP) and DPPH assays. In the context of FRAP, ethanol of riceberry broken rice extract showed the highest FRAP amount (1.33 mg $FeSO_4$/g extract) compared to ethyl acetate, ethanol and water, and water at 0.26, 0.55, and 1.19 mg $FeSO_4$/g extract (Luang-In et al. 2018). In another study, ethanolic riceberry rice extract exhibited an effective FRAP value at 1.64 $FeSO_4$ μg/mL (Soradech et al. 2016), which is higher than that of reported by Luang-In et al. (2018). Water and methanol extracts of brewers' rice exhibited 4.77–4.88 mg Fe/g of FRAP (Tan et al. 2013). The amount of FRAP in brewers' rice was higher than that found in different cultivars of Thailand rice bran (0.10–0.53 mg/mL) (Chotimarkorn et al. 2008). The reducing power of rice bran extract was linked to the total phenolic content (Chotimarkorn et al. 2008). Notably, the antioxidant activity and phenolic content of stabilized rice bran were increased when the infrared radiation power was increased (Irakli et al. 2018).

The data further revealed that total phenolic content in ethanol and water, water, and ethanol of riceberry broken rice (7.24–9.94 mg GAE/g) was three-fold higher than that found in rice bran of several rice varieties in Thailand (2.2–3.2 mg GAE/g) (Chotimarkorn et al. 2008). Intriguingly, the total phenolic content in riceberry broken rice was 40-fold higher than brewers' rice (0.20–0.32 mg GAE/g) (Tan et al. 2013). Further, the total phenolic content in riceberry broken rice was ten-fold higher than KDML105 jasmine broken rice extract (0.121–0.127 mg GAE/g) (Kanpai and Thitipramote 2015), implied that pigmented riceberry broken rice is rich in phenolic compounds compared to jasmine broken rice. Moongngarm et al. (2012) evaluated the total antioxidant capacity of rice germ, rice bran layer (without germ), and rice bran (contained rice germ and bran layer). Among the rice germ, rice bran layer, and rice bran, rice germ of black rice contained the highest total antioxidant capacity. While white rice (RD6 and KDML) has comparatively lower total antioxidant capacity. In this regard, the rice bran and rice germ displayed electron-donating ability and thereby may serve as a radical chain terminator to produce more stable products from reactive free radical species (Arabshshi and Urooj 2007).

It has been demonstrated that rice husk is a valuable source of bioactive components containing high antioxidant properties. Rice husk contains phenolic acid that can prevent rice seed from oxidative stress (Butsat and Siriamornpun 2010). In this regard, it has been identified as a potential source of organic chemicals and energy (Moure et al. 2001; Mochidzuki et al. 2001). Treatment with far-infrared radiation for 30 min enhanced the radical scavenging activity and total phenolic content of rice husk. Far-infrared radiation onto rice husk was also inhibited the lipid peroxidation and produced more phenolic compounds. This finding implied that far-infrared radiation onto rice husk may activate and liberate covalently bound phenolic compounds (Lee et al. 2003). The previous study stated that hydrothermal treatment of rice husk produces lignin-derived compounds, for instance, ferulic and caffeic acids (Garrote et al. 2007). These components are important for the pharmaceutical industry due to their protection against photooxidative damage (Saija et al. 1999). The phenolic compounds in the methanolic extract of rice husk have demonstrated high antioxidant activity by scavenging hydrogen peroxide-induced damage and singlet oxygen in human lymphocytes (Jeon et al. 2006). Phenolic compounds have been recognized to exert potent antioxidant activity (Zieliński and Kozłowska 2000; Heim et al. 2002; Rastija and Medić-Šarić 2009). Indeed, the biological activity of the cereal grains was linked to their polyphenolic compounds (Awika et al. 2003). In a study reported by Li et al. (2008) revealed that most of the phenolic acids in the whole grains are syringic, caffeic, vanillic, p-coumaric, and ferulic acids. Another rice by-product, rice straw is rich in phenolic compounds including sinapic, p-coumaric, ferulic, and vanillic acids as well as catechin and flavonoids aglycones (kaempferol, apigenin, and quercetin). In particular, ferulic and p-coumaric acids are the major phenolic acids identified in rice straw followed by sinapic and vanillic acids, contributing 15% and 19%, 2.4% and 2%, respectively. By contrast, flavonoidal aglycones (kaempferol, apigenin, and quercetin) and catechin recorded 5%, 3%, 7%, and 4%, respectively (Meselhy et al. 2019).

Phenolics are components exert one or more aromatic rings with hydroxyl groups, and commonly classified as tannins, coumarins, flavonoids, stilbenes, and phenolic acids (Luna-Guevara et al. 2018). Phenolic compounds are negatively linked to the risk of chronic diseases (Luna-Guevara et al. 2018). Several studies have demonstrated that polyphenols exert anti-inflammatory properties, modulate cellular signaling, and enhance endothelial function (Chew et al. 2019). It remains unknown whether other mechanism or its antioxidant is responsible for its protective effect, evidence strongly supports an inverse association between polyphenol intake and risk of certain metabolic ailments (Corrêa and Rogero 2019; Sut et al. 2019). The ability to regulate apoptotic cells has been demonstrated as a potential therapeutic approach for cancer. Many bioactive constituents present in rice bran such as p-coumaric acid (Janicke et al. 2005) and ferulic acid (Huang and Ng 2012) are believed to be responsible for the induction of apoptotic cell death (Ryan et al. 2011). Ferulic acid and caffeic acid derivatives play a critical role in elevating of tumor suppressor protein p53 levels and improving the chromatin condensation and mitochondrial depolarization (Serafim et al. 2011) (Fig. 4.8).

Dietary phenolic acids possess physiological antioxidant properties by quenching ROS and reactive nitrogen species (RNS). Hence, it is potentially modifying the pathogenic mechanisms related to CVD. ROS and RNS damage the cells, possibly leading to cellular dysfunction and disease (Allan Butterfield and Halliwell 2019; Singh et al. 2019). For example, oxidative damage to lipoproteins, especially LDL, can generate modified particles containing damaged apoprotein and lipid oxidation products, in which the oxidized LDL may enhance the atherogenic effects (Kattoor et al. 2019). Oxidation is believed to occur in the subendothelial space of the arterial wall (Hartley et al. 2019). It has been suggested that oxidized LDL contributes to all stages of atherosclerotic process, such as increase uptake of oxidized LDL to form foam cells, recruitment of macrophages, promote endothelial damage, and stimulation of inflammatory response (Zhong et al. 2019). The previous study has demonstrated that redox-active phenolic compounds can serve as a pro-oxidant under certain circumstances. This is commonly found in the metal ion catalyzed system, as evaluated with flavonoids (González-Paramás et al. 2019). Several studies showed that a diet containing high amounts of flavonoid is inversely related to CVD. Flavonoids (Fig. 4.5) prevent CVD through different mechanisms including antiplatelet, anti-inflammatory, antioxidant, and increasing HDL-C levels (Nunes et al. 2016) (Fig. 4.8). Furthermore, caffeic acid has also been demonstrated to have pro-oxidant effects in Cu^{2+}-induced oxidation of LDL (Barnaba and

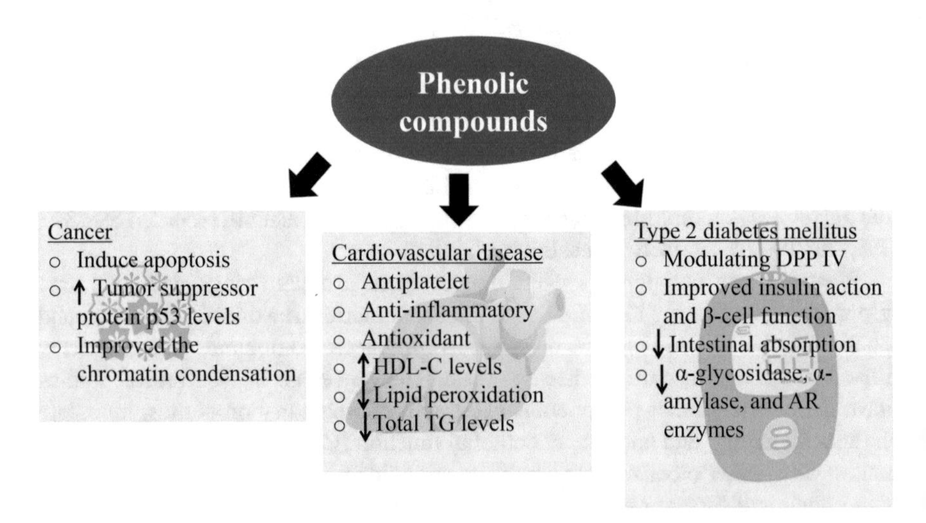

Fig. 4.8 Effect of phenolic compounds on cancer, cardiovascular disease, and type 2 diabetes mellitus. *AR* aldose reductase, *DPP IV* dipeptidyl peptidase IV, *HDL-C* high-density lipoprotein cholesterol, *TG* triglycerides

Medina-Meza 2019). Pro-oxidant activity is often observed in the propagation phase of oxidation. Lai et al. (2015) showed that dietary consumption of fruits, rich in phenolic compounds, were linked to the improvement of cardiovascular health by reducing nearly 6–7% deaths from CVD. An animal study further demonstrated that feeding mice with 0.5 μmol gallic acid/day (date seed, Hallawi extracts, and pomegranate juice for 3 weeks) reduced lipid peroxidation, oxidative stress, and total triglycerides (TG) levels in macrophages in the aortas, suggesting that combination of polyphenols from different sources enhanced the reduction of oxidative stress and lipids (Rosenblat et al. 2015). This finding suggests that dietary flavonoids or phenolic compounds could serve as a natural cardiovascular protector.

Diabetic mellitus is suffered by older people and often resulting in severe complications including diabetic peripheral neuropathy (Tesfaye and Wu 2018). Dipeptidyl peptidase IV (DPP IV) has been recognized as an effective target in the management of blood glucose by inhibiting DPP IV (Barnett 2006). Natural phenolic compounds ameliorate blood glucose in diet-induced diabetic ICR mice and S961-treated C57BL/6 mice by targeting DPP IV (Huang et al. 2019). DPP IV is a membrane surface antigen protein or known as adenosine deaminase complexing protein 2, which is involved in apoptosis-associated proteins, signal transduction, and immune system (Pro and Dang 2004). Dietary phenolic compounds have shown an anti-diabetic potential by delay or prevent the development of diabetes mellitus (Babu et al. 2013). Phenolic compounds modulate diabetes mellitus through various mechanisms such as improves insulin action and β-cell function, modulates enzymes involved in the glucose metabolism, and decreases intestinal absorption and dietary carbohydrates digestion (Jung et al. 2004; Iwai et al. 2006; Iwai 2008). α-glycosidase, α-amylase, and aldose reductase (AR) is a crucial enzyme to prevent diabetic complications. It has been demonstrated that phenolic compounds inhibit α-glycosidase, α-amylase, and AR enzymes (Demir et al. 2019) (Fig. 4.8). Together, polyphenolic compounds show promising effects in several human diseases. Intakes of food rich in phenolic compounds promote the effectiveness of these compounds through synergism effects. However, further studies are needed to assess the safety and adequate doses of phenolic compounds against diseases. Collectively, the extraction time and solvents may play a critical role in extracting bioactive compounds. More studies are warranted to assess the optimal extraction conditions on a large scale of bioactive compounds for industrial purposes.

4.6 Other Potential Component in Rice By-products

4.6.1 Dietary Fiber

Dietary fiber is an analogous carbohydrate or edible part of plants that are resistant to digestion and absorption in the human small intestine with partial fermentation in the large intestine (Prasad and Bondy 2019). Although dietary fiber is a plant food material that is not hydrolyzed by enzyme secreted from the human digestive tracts, the microflora in the gut may facilitate the digestion. These plant food materials include non-starch polysaccharides, for example, pectins, gums, some hemicelluloses, and celluloses (Ciudad-Mulero et al. 2019). In general, dietary fiber can be divided into two classes depending on the water solubility (Dai and Chau 2017). The matrix or structural fibers such as some hemicelluloses, lignins, and celluloses are insoluble, while natural gel-forming fibers, for instance, mucilage, gums, inulin, resistant starch, fructo-oligosaccharide, and pectin are soluble (Palamae et al. 2017) (Fig. 4.9). The composition of dietary fiber in rice straw, bran, and husk are shown in Table 4.3.

Some of the previous studies found that fiber content in rice germ (9.52%) of rice cultivar Khao dok mali 105 was much lower than rice bran (12.48%) (Moongngarm et al. 2012). The study further revealed that the amount of fiber in rice germ (11.67%) was almost similar to rice bran (11.77%) in rice cultivar RD6. In the context of pigmented rice, rice germ of black rice and red rice contains higher levels of fiber (14.07–17.42%) compared to rice bran (12.11–12.68%) (Moongngarm et al. 2012). Fadaei and Salehifar (2012) reported that enzymatic extraction of rice husk has more than 60% dietary fiber, which is higher than chemical extraction (44.66 g/100 g). During enzymatic treatment, rice husk is treated with protease, amyloglucosidase, and amylase to remove other impurities, protein, and starch, and thereby producing higher dietary fiber (Fadaei and Salehifar 2012). In addition, rice husk fibers extracted by the enzyme (7.00 mL/g) have stronger water-binding capacity compared to chemical extraction (6.03 mL/g) (Fadaei and Salehifar 2012). The number of hydroxyl group presents in the fiber may contribute to the water absorption and thus allow the water interaction via hydrogen bonding (Ferreira et al. 2017). Water plays a crucial role in baking including color and flavor formation, enzyme and

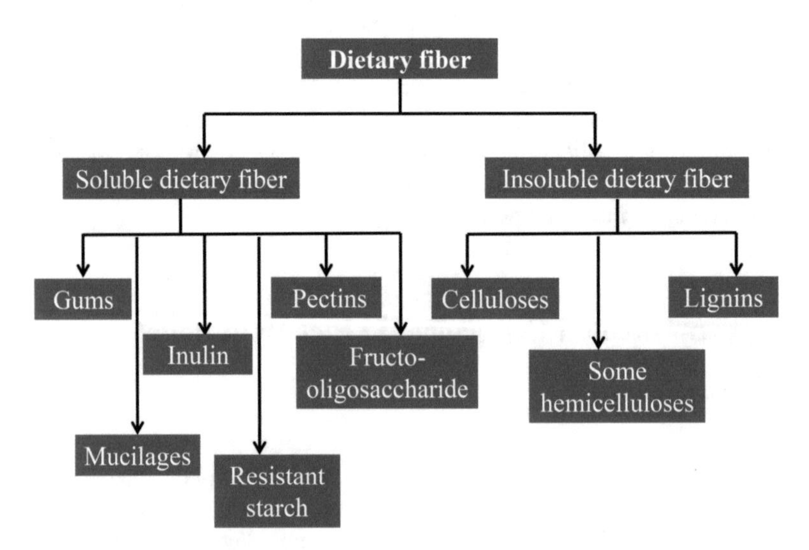

Fig. 4.9 Classification of dietary fiber according to chemical properties

Table 4.3 The composition of dietary fiber in rice straw, bran, and husk

Dietary fiber (%)	Rice husk[a]	Rice bran[a]	Rice straw[b]
Cellulose	38	30	32.0
Hemicellulose	20	20	35.7
Lignin	22	20	22.3

[a]Organisation for Economic Co-operation and Development (2004)
[b]Worasuwannarak et al. (2007)

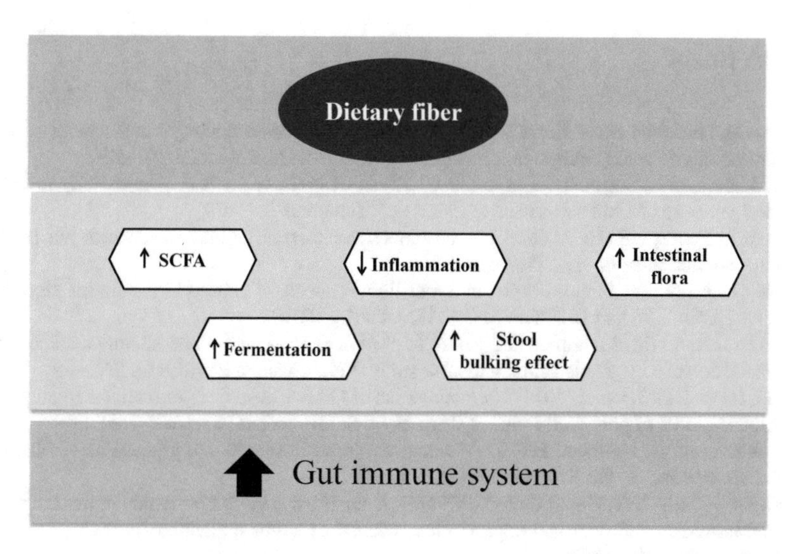

Fig. 4.10 Dietary fiber enhances gut immune system via a few mechanisms. *SCFA* short-chain fatty acids

yeast inactivation, denaturation, and gelatinization (Schiraldi and Fessas 2012). It has been demonstrated that addition dietary fiber to bread increased bread firmness and decreased volume of bread (Kurek and Wyrwisz 2015).

Dietary fibers play a crucial role in enhancing the immune system via the production of short-chain fatty acids (SCFA), implies that increase consumption of fermentable dietary fibers may be crucial in the reduction of inflammation (Maslowski et al. 2009; Silveira et al. 2017) (Fig. 4.10). SCFA may increase macrophages, T helper cells, neutrophils, and cytotoxic activity in NK cells (Slavin 2013). Additionally, the changes in gut microbiota and fermentation of dietary fiber are linked to impaired gastrointestinal tolerance (Hur and Lee 2015). Colonic and mucosal microflora prevents pathogenic bacteria from invading the gastrointestinal tract that facilitates with the gut immune system (Kamada and Núñez 2014). Fermentation of undigested carbohydrates in the upper gut promotes the intestinal flora salvage energy (Slavin 2013). The main substrates are mucus and dietary carbohydrates, which escape digestion in the upper gastrointestinal tract (Slavin 2013). The production of pyruvate from hexoses in undigested carbohydrates is primarily through the fermentation pathway (Slavin 2013). Colonic bacteria use different carbohydrates to hydrolyze enzymes and produce lactate, carbon dioxide, hydrogen, methane, and SCFA (predominantly acetate, propionate, and butyrate) (Vogt et al. 2015). Therefore, these compounds enhance fecal and bacterial mass, increase fermentation, and subsequently lead to a stool bulking effect (Slavin 2013). Given its high nutritional values, it would be of great interest in future studies to explore ways for rice by-products to be incorporated into nutritious foods.

References

Abdou AM, Higashiguchi S, Horie K et al (2006) Relaxation and immunity enhancement effects of γ-aminobutyric acid (GABA) administration in humans. Biofactors 26:201–208

Abe Y, Umemura S, Sugimoto K et al (1995) Effect of green tea rich in γ-aminobutyric acid on blood pressure of Dahl salt-sensitive rats. Am J Hypertens 8:74–79

Adeghate E, Ponery AS (2002) GABA in the endocrine pancreas: cellular localization and function in normal and diabetic rats. Tissue Cell 34:1–6

AgrEvo (1999) Safety, compositional and nutritional aspects of LibertyLink rice transformation events LLRICE06 and LLRICE62. U.S. FDA/CFSANBNF 63

Ahmed S, Hasan MM, Mahmood ZA (2016) Inhibition and modulation of calcium oxalate monohydrate crystals by phytic acid: an *in vitro* study. J Pharmacogn Phytochem 5:91–95

Ahn JH, Im C, Park JH et al (2014) Hypnotic effect of GABA from rice germ and/or tryptophan in a mouse model of pentothal-induced sleep. Food Sci Biotechnol 23:1683–1688

Allan Butterfield D, Halliwell B (2019) Oxidative stress, dysfunctional glucose metabolism and Alzheimer disease. Nat Rev Neurosci 20:148–160

Ambrogini P, Minelli A, Galati C et al (2014) Post-seizure α-tocopherol treatment decreases neuroinflammation and neuronal degeneration induced by status epilepticus in rat hippocampus. Mol Neurobiol 50:246–256

AOAC (Association of Official Analytical Chemists) (2002) Official methods of analysis of AOAC international, 17th edn. AOAC, Gaithersburg, MD. Chapter 4, pp 20–27

Arabshshi S, Urooj A (2007) Antioxidant properties of various solvent extracts of mulberry (Morus indica L.) leaves. Food Chem 102:1233–1240

Awika JM, Dykes L, Gu L et al (2003) Processing of sorghum (Sorghum bicolor) and sorghum products alters procyanidin oligomer and polymer distribution and content. J Agric Food Chem 51:5516–5521

Babu PVA, Liu D, Gilbert ER (2013) Recent advances in understanding the anti-diabetic actions of dietary flavonoids. J Nutr Biochem 24:1777–1789

Bakir DS, Yalcin G, Cucu AK (2020) Isolation and determination of tocopherols and tocotrienols from the seed of *Capparis Ovata* grown in Turkey by reversed-phase high-performance liquid chromatography. Chromatographia 83:77–86

Barker CJ, Berggren PO (1999) Inositol hexakisphosphate and beta-cell stimulus-secretion coupling. Anticancer Res 19:3737–3741

Barnaba C, Medina-Meza IG (2019) Flavonoids ability to disrupt inflammation mediated by lipid and cholesterol oxidation. In: Honn K, Zeldin D (eds) The role of bioactive lipids in cancer, inflammation and related diseases. Adv Exp Med Biol 1161:243–253

Barnett A (2006) DPP-4 inhibitors and their potential role in the management of type 2 diabetes. Int J Clin Pract 60:1454–1470

Battelli MG, Corte ED, Stirpe F (1972) Xanthine oxidase type D (dehydrogenase) in the intestine and other organs of the rat. Biochem J 126:747–749

Bhagwagar Z, Wylezinska M, Jezzard P et al (2007) Reduction in occipital cortex γ-aminobutyric acid concentrations in medication-free recovered unipolar depressed and bipolar subject. Biol Psychiatry 61:806–812

Bohn L, Meyer AS, Rasmussen SK (2008) Phytate: impact on environment and human nutrition. A challenge for molecular breeding. J Zhejiang Univ Sci B 9:165–191

Braun M, Ramracheya R, Bengtsson M et al (2010) γ-Aminobutyric acid (GABA) is an autocrine excitatory transmitter in human pancreatic β-cells. Diabetes 59:1694–1701

Braun M, Wendt A, Birnir B et al (2004) Regulated exocytosis of GABA-containing synaptic-like microvesicles in pancreatic β-cells. J Gen Physiol 123:191–204

Budin SB, Othman F, Louis SR et al (2009) The effects of palm oil tocotrienol-rich fraction supplementation on biochemical parameters, oxidative stress and the vascular wall of streptozotocin-induced diabetic rats. Clin (São Paulo, Brazil) 64:235–244

Butsat S, Siriamornpun S (2010) Antioxidant capacities and phenolic compounds of the husk, bran and endosperm of Thai rice. Food Chem 119:606–613

Butterfield DA, Boyd-Kimball D (2005) The critical role of methionine 35 in Alzheimer's amyloid beta-peptide (1-42)-induced oxidative stress and neurotoxicity. Biochim Biophys Acta 1703:149–156

Chen L, Zhao H, Zhang C et al (2016) γ-Aminobutyric acid-rich yogurt fermented by *Streptococcus salivarius* subsp. *thermophiles* fmb5 apprars to have anti-diabetic effect on streptozotocin-induced diabetic mice. J Funct Foods 20:267–275

Chew B, Mathison B, Kimble L et al (2019) Chronic consumption of a low calorie, high polyphenol cranberry beverage attenuates inflammation and improves glucoregulation and HDL cholesterol in healthy overweight humans: a randomized controlled trial. Eur J Nutr 58:1223–1235

Chotimarkorn C, Benjakul S, Silalai N (2008) Antioxidant components and properties of five long-grained rice bran extracts from commercial available cultivars in Thailand. Food Chem 111:636–641

Cicero AF, Gaddi A (2001) Rice bran oil and gamma-oryzanol in the treatment of hyperlipoproteinaemias and other conditions. Phytother Res 15:277–289

Ciudad-Mulero M, Fernández-Ruiz V, Matallana-González MC et al (2019) Dietary fiber sources and human benefits: the case study of cereal and pseudocereals. Adv Food Nutr Res 90:83–134

Corrêa TAF, Rogero MM (2019) Polyphenols regulating microRNAs and inflammation biomarkers in obesity. Nutrition 59:150–157

Coupland NJ, Nutt DJ (1995) Neurobiology of anxiety and panic. In: Bradwein J, Vasar E (eds) Cholecystokinin and anxiety: from neuron to behaviour. Springer, New York, pp 1–31

Da Silva EO, Gerez JR, Hohmann MSN et al (2019) Phytic acid decreases oxidative stress and intestinal lesions induced by fumonisin B1 and deoxynivalenol in intestinal explants of pigs. Toxins (Basel) 11:18

Dai F-J, Chau C-F (2017) Classification and regulatory perspectives of dietary fiber. J Food Drug Anal 25:37–42

Dargaei Z, Bang JY, Mahadevan V et al (2018) Restoring GABAergic inhibition rescues memory deficits in a Huntington's disease mouse model. Proc Natl Acad Sci USA 115:E1618–E1626

Demir Y, Durmaz L, Taslimi P et al (2019) Antidiabetic properties of dietary phenolic compounds: Inhibition effects on α-amylase, aldose reductase, and α-glycosidase. Biotechnol Appl Biochem 66:781–786

Devaraj S, Jialal I (1998) The effects of alpha-tocopherol on critical cells in atherogenesis. Curr Opin Lipidol 9:11–15

Dhakal R, Bajpai VK, Baek K (2012) Production of GABA (γ-Aminobutyric acid) by microorganisms: a review. Braz J Microbiol 43:1230–1241

Diana M, Quílez J, Rafecas M (2014) γ-Aminobutyric acid as a bioactive compound in foods: a review. J Funct Foods 10:407–420

Dou H, Wang C, Wu X et al (2014) Calcium influx activates adenylyl cyclase 8 for sustained insulin secretion in rat pancreatic beta cells. Diabetologia 58:324–333

Drake DJ, Nader G, Forero L (2002) Feeding rice straw to cattle. Publication 8079. University of California, Division of Agriculture and Natural Resources. http://anrcatalog.ucdavis.edu
Esa NM, Ling TB, Peng LS (2013) By-products of rice processing: an overview of health benefits and applications. J Rice Res 1:107
Fadaei V, Salehifar M (2012) Rice husk as a source of dietary fiber. Ann Biol Res 3:1437–1442
Fadel JG, MacKill DJ (2002) Characterization of rice straw—94. U. of California, Davis, CA. http://www.syix.com/rrb/94rpt/RiceStraw.htm
FAO (2003) Animal feed resources information system. *Oryza sativa*. http://www.fao.org/livestock/agap/frg/AFRIS/Data/312.htm
Farrell DJ, Hutton K (1990) Chapter 24. Rice and rice milling by-products. In: Thacker, Kirkwood (eds) Nontraditional feed sources for use in swine production. Butterworths Publishers, Stoneham
Ferreira SR, de Andrade SF, Lima PRL et al (2017) Effect of hornification on the structure, tensile behavior and fiber matrix bond of sisal, jute and curauá fiber cement based composite systems. Construct Build Mater 139:551–561
Ffoulkes D (1998) Rice as a livestock feed. Agnote (Northern Territory of Australia). https://transact.nt.gov.au/ebiz/dbird/TechPublications.nsf/.../273.pdf
Forster GM, Raina K, Kumar A et al (2013) Rice varietal differences in bioactive bran components for inhibition of colorectal cancer cell growth. Food Chem 141:1545–1552
Galli F, Azzi A, Birringer M et al (2017) Vitamin E: emerging aspects and new directions. Free Radic Biol Med 102:16–36
Garcia-Estepa RM, Guerra-Hernandez E, Garcia-Villanova B (1999) Phytic acid content in milled cereal products and breads. Food Res Int 32:217–221
Garrote G, Falqué E, Domínguez H et al (2007) Autohydrolysis of agricultural residues: study of reaction byproducts. Bioresour Technol 98:1951–1957
González-Paramás AM, Ayuda-Durán B, Martínez S et al (2019) The mechanisms behind the biological activity of flavonoids. Curr Med Chem 26:6976–6990
Graf E, Eaton JW (1990) Antioxidant functions of phytic acid. Free Radic Biol Med 8:61–69
Graf E, Empson KL, Eaton JW (1987) Phytic acid. A natural antioxidant. J Biol Chem 262:11647–11650
Gray JA, Quintero S, Mellanby J et al (1984) Some biochemical, behavioural and electrophysiological tests of the GABA hypothesis of anti-anxiety drug action. In: Bowery N (ed) Actions and interactions of GABA and benzodiazepines. Raven Press, New York, pp 237–262
Guzmán BC-F, Vinnakota C, Govindpani K et al (2018) The GABAergic system as a therapeutic target for Alzheimer's disease. J Neurochem 146:649–669
Hartley A, Haskard D, Khamis R (2019) Oxidized LDL and anti-oxidized LDL antibodies in atherosclerosis–novel insights and future directions in diagnosis and therapy. Trends Cardiovasc Med 29:22–26
Hasler G, van der Veen JW, Tumonis T et al (2007) Reduced prefrontal glutamate/glutamine and γ-aminobutyric acid levels in major depression determined using proton magnetic resonance spectroscopy. Arch Gen Psychiatry 64:193–200
Hayakawa K, Kimura M, Kasaha K et al (2004) Effect of a γ-aminobutyric acid-enriched dairy product on the blood pressure of spontaneously hypertensive and normotensive WistarKyoto rats. Br J Nutr 92:411–417
Heim KE, Tagliaferro AR, Bobilya DJ (2002) Flavonoid antioxidants: chemistry, metabolism and structure-activity relationships. J Nutr Biochem 13:572–584
Heinemann RJ, Xu Z, Godber JS et al (2008) Tocopherols, tocotrienols, and γ-oryzanol contents in japonica and indica subspecies of rice (*Oryza sativa* L.) cultivated in Brazil. Cereal Chem 85:243–247
Herd D (2003) Composition of alternative feeds—dry basis. Texas A&M U, College Station. http://animalscience.tamu.edu/ansc/publications/beefpubs/asw012-altfeeds.pdf
Holifa A, Latif AZA, Simbak NB et al (2017) Alpha-tocopherol administration in diabetics as preventive and therapeutic agents in oxidative stress. Curr Trends Biomed Eng Biosci 5:555671

Huang P-K, Lin S-R, Chang C-H et al (2019) Natural phenolic compounds potentiate hypoglycemia via inhibition of dipeptidyl peptidase IV. Sci Rep 9:15585
Huang S-H, Ng L-T (2012) Quantification of polyphenolic content and bioactive constituents of some commercial rice varieties in Taiwan. J Food Compos Anal 26:122–127
Hur KY, Lee M-S (2015) Gut microbiota and metabolic disorders. Diabetes Metab J 39:198–203
Imsanguan P, Roaysubtawee A, Borirak R et al (2008) Extraction of α-tocopherol and γ-oryzanol from rice bran. LWT Food Sci Technol 41:1417–1424
Inoue K, Shirai T, Ochiai H et al (2003) Blood-pressure-lowering effect of a novel fermented milk containing γ-aminobutyric acid (GABA) in mild hypertensives. Eur J Clin Nutr 57:490–495
Irakli M, Kleisiaris F, Mygdalia A et al (2018) Stabilization of rice bran and its effect on bioactive compounds content, antioxidant activity and storage stability during infrared radiation heating. J Cereal Sci 80:135–142
Islam MS, Murata T, Fujisawa M et al (2008) Anti-inflammatory effects of phytosteryl ferulates in colitis induced by dextran sulphate sodium in mice. Br J Pharmacol 154:812–824
Islam MS, Yoshida H, Matsuki N et al (2009) Antioxidant, free radical-scavenging, and NF-kappaB-inhibitory activities of phytosteryl ferulates: structure-activity studies. J Pharmacol Sci 111:328–337
Islam S, Nagasaka R, Ohara K et al (2011) Biological abilities of rice bran-derived antioxidant phytochemicals for medical therapy. Curr Top Med Chem 11:1847–1853
Iwai K (2008) Antidiabetic and antioxidant effects of polyphenols in brown alga Ecklonia stolonifera in genetically diabetic KK-Ay mice. Plant Foods Hum Nutr 63:163–169
Iwai K, Kim M-Y, Onodera A et al (2006) α-glucosidase inhibitory and antihyperglycemic effects of polyphenols in the fruit of viburnum dilatatum thunb. J Agric Food Chem 54:4588–4592
Janicke B, Önning G, Oredsson SM (2005) Differential effects of ferulic acid and p-coumaric acid on S phase distribution and length of S phase in the human colonic cell line Caco-2. J Agric Food Chem 53:6658–6665
Jeon K-I, Park E, Park H-R et al (2006) Antioxidant activity of far-infrared radiated rice hull extracts on reactive oxygen species scavenging and oxidative DNA damage in human lymphocytes. J Med Food 9:42–48
Juliano BO, Bechtel DB (1985) The rice grain and its gross composition. In: Juliano BO (ed) Rice: chemistry and technology. American Association of Cereal Chemists, St. Paul, MN, pp 17–57
Juliano C, Cossu M, Alamanni MC et al (2005) Antioxidant activity of gamma-oryzanol: mechanism of action and its effect on oxidative stability of pharmaceutical oils. Int J Pharm 299:146–154
Jung UJ, Lee M-K, Jeong K-S et al (2004) The hypoglycemic effects of hesperidin and naringin are partly mediated by hepatic glucose-regulating enzymes in C57BL/KsJ-db/db mice. J Nutr 134:2499–2503
Kamada N, Núñez G (2014) Regulation of the immune system by the resident intestinal bacteria. Gastroenterology 146:1477–1488
Kanpai S, Thitipramote N (2015) Development of standardized jasmine rice 105 for antioxidant application. MSc Thesis in Science. Mae Fah Luang University, Thailand
Kartikawati M, Purnomo H (2019) Improving meatball quality using different varieties of rice bran as natural antioxidant. Food Res 3:79–85
Kasim AB, Edwards HM (1998) The analysis for inositol phosphate forms in feed ingredients. J Sci Food Agric 76:1–9
Kattoor AJ, Kanuri SH, Mehta JL (2019) Role of Ox-LDL and LOX-1 in atherogenesis. Curr Med Chem 26:1693–1700
Khanna S, Parinandi NL, Kotha SR et al (2010) Nanomolar vitamin E α-tocotrienol inhibits glutamate-induced activation of phospholipase A2 and causes neuroprotection. J Neurochem 112:1249–1260
Knuckles BE, Kuzmicky DD, Betschart AA (1982) HPLC analysis of phytic acid in selected foods and biological samples. J Food Sci 47:1257–1262

Kook M-C, Seo M-J, Cheigh C-I et al (2010) Enhanced production of γ-aminobutyric acid using rice bran extracts by *Lactobacillus sakei* B2-16. J Microbiol Biotechnol 20:763–766
Kurek M, Wyrwisz J (2015) The application of dietary fiber in bread products. J Food Process Technol 6:447
Lacerda JEC, Campos RR, Araujo GC et al (2003) Cardiovascular responses to microinjections of GABA or anesthetics into the rostral ventrolateral medulla of conscious and anesthetized rats. Braz J Med Biol Res 36:1269–1277
Lai HT, Threapleton DE, Day AJ et al (2015) Fruit intake and cardiovascular disease mortality in the UK Women's Cohort Study. Eur J Epidemiol 30:1035–1048
Lai P, Li KY, Lu S et al (2009) Phytochemicals and antioxidant properties of solvent extracts from Japonica rice bran. Food Chem 117:538–544
Larsson O, Barker CJ, Sjöholm A et al (1997) Inhibition of phosphatases and increased Ca2+ channel activity by inositol hexakisphosphate. Science 278:471–474
Lee S, Yu S, Park HJ et al (2019) Rice bran oil ameliorates inflammatory responses by enhancing mitochondrial respiration in murine macrophages. PLoS One 14:e0222857
Lee S-C, Kim J-H, Jeong S-M et al (2003) Effect of far-infrared radiation on the antioxidant activity of rice hulls. J Agric Food Chem 51:4400–4403
Li H, Cao Y (2010) Lactic acid bacterial cell factories for gamma-aminobutyric acid. Amino Acids 39:1107–1116
Li L, Fu Q, Xia M et al (2018) Inhibition of P-glycoprotein mediated efflux in Caco-2 cells by phytic acid. J Agric Food Chem 66:988–998
Li L, Shewry PR, Ward JL (2008) Phenolic acids in wheat varieties in the HEALTHGRAIN diversity screen. J Agric Food Chem 56:9732–9739
Ligon B, Yang J, Morin SB et al (2007) Regulation of pancreatic islet cell survival and replication by γ-aminobutyric acid. Diabetologia 50:764–773
Limtrakul P, Yodkeeree S, Pitchakarn P et al (2015) Suppression of inflammatory responses by black rice extract in RAW 264.7 macrophage cells via downregulation of NF-kB and AP-1 signaling pathways. Asian Pac J Cancer Prev 16:4277–4283
Limtrakul P, Yodkeeree S, Pitchakarn P et al (2016) Anti-inflammatory effects of proanthocyanidin-rich red rice extract via suppression of MAPK, AP-1 and NF-κB pathways in raw 264.7 macrophages. Nutr Res Pract 10:251–258
Lee LS, Lee N, Kim YH et al (2013) Optimization of ultrasonic extraction of phenolic antioxidants from green tea using response surface methodology. Molecules 18:13530–13545
Lopez HW, Leenhardt F, Coudray C (2002) Minerals and phytic acid interactions: is it a real problem for human nutrition? Int J Food Sci Technol 37:727–739
Luang-In V, Yotchaisarn M, Somboonwatthanakul I et al (2018) Bioactivities of organic riceberry broken rice and crude riceberry rice oil. Thai J Pharm Sci 42:161–168
Luna-Guevara ML, Luna-Guevara JJ, Hernández-Carranza P et al (2018) Chapter 3—Phenolic compounds: a good choice against chronic degenerative diseases. Stud Nat Prod Chem 59:79–108
Mangialasche F, Solomon A, Kareholt I et al (2013a) Serum levels of vitamin E forms and risk of cognitive impairment in a Finnish cohort of older adults. Exp Gerontol 48:1428–1435
Mangialasche F, Westman E, Kivipelto M et al (2013b) Classification and prediction of clinical diagnosis of Alzheimer's disease based on MRI and plasma measures of alpha−/gamma-tocotrienols and gamma-tocopherol. J Intern Med 273:602–621
Maslowski KM, Vieira AT, Ng A et al (2009) Regulation of inflammatory responses by gut microbiota and chemoattractant receptor GPR43. Nature 461:1282–1286
Meselhy KM, Shams MM, Sherif NH et al (2019) Phenolic profile and in vivo cytotoxic activity of rice straw extract. Pharmacogn J 11:849–857
Miller ER, Ullrey DE, Lewis AJ (1991) Swine nutrition. Butterworth–Heinemann, Boston/London
Mochidzuki K, Sakoda A, Suzuki M et al (2001) Structural behavior of rice husk silica in pressurized hot-water treatment processes. Ind Eng Chem Res 40:5705–5709

Moongngarm A, Daomukda N, Khumpika S (2012) Phytochemicals, and antioxidant capacity of rice bran, rice bran layer, and rice germ. APCBEE Proc 2:73–79
Moure A, Cruz JM, Franco D et al (2001) Natural antioxidants from residual sources. Food Chem 72:145–171
Muraoka S, Miura T (2004) Inhibition of xanthine oxidase by phytic acid and itsantioxidative action. Life Sci 74:1691–1700
Naito Y, Shimozawa M, Kuroda M et al (2005) Tocotrienols reduce 25-hydroxycholesterol-induced monocyte-endothelial cell interaction by inhibiting the surface expression of adhesion molecules. Atherosclerosis 180:19–25
Nam SH, Choi SP, Kang MY et al (2006) Antioxidative activities of bran extracts from twenty one pigmented rice cultivars. Food Chem 94:613–620
Nayak P, Sharma SB, Chowdary NVS (2019) Alpha-tocopherol supplementation restricts aluminium- and ethanol-induced oxidative damage in rat brain but fails to protect against neurobehavioral damage. J Diet Suppl 16:257–268
Ng LT, Ko HJ (2012) Comparative effects of tocotrienol-rich fraction, α-tocopherol and α-tocopheryl acetate on inflammatory mediators and nuclear factor kappa B expression in mouse peritoneal macrophages. Food Chem 134:920–925
NGFA (National Grain & Feed Association) (2003) Rice fractions. NGFA, Grafton. http://www.Ingredients101.com/ricefrac.htm
Niki E (2014) Role of vitamin E as a lipid-soluble peroxyl radical scavenger: in vitro and in vivo evidence. Free Radic Biol Med 66:3–12
Nour AM (2003) Rice straw and rice hulls in feeding ruminants in Egypt. Department of Animal Production, Alexandria University, Alexandria. http://www.ssdairy.org/AdditionalRes/x5494e/x5494e07.htm
NRC (1994) Nutrient requirements of poultry, 9th revised edn. National Academy Press, Washington DC
NRC (1998) Nutrient requirements of swine, 10th revised edn. National Academy Press, Washington DC. http://books.nap.edu/catalog/2114.html
NRC (2000) Nutrient requirements of beef cattle (Update 2000). National Academy Press, Washington DC
NRC (2001) Nutrient requirements of dairy cattle, 7th revised edn. National Academy Press, Washington DC. http://books.nap.edu/catalog/9825.html
NRC (National Research Council) (1982) United States–Canadian tables of feed composition. National Academy Press, Washington DC
Nunes MA, Pimentel F, Costa ASG et al (2016) Cardioprotective properties of grape seed proanthocyanidins: an update. Trends Food Sci Technol 57:31–39
Organisation for Economic Co-operation and Development (2004) Consensus document on compositional considerations for new varieties of rice (*Oryza sativa*): key food and feed nutrients and anti-nutrients. In: Annual report 2004. Paris, 17 August
Özer NK, Palozza P, Boscoboinik D et al (1993) D-alpha-Tocopherol inhibits low density lipoprotein induced proliferation and protein kinase C activity in vascular smooth muscle cells. FEBS Lett 322:307–310
Özer NK, Sirikci O, Taha S et al (1998) Effect of vitamin E and probucol on dietary cholesterol-induced atherosclerosis in rabbits. Free Radic Biol Med 24:226–233
Palamae S, Dechatiwongse P, Choorit W et al (2017) Cellulose and hemicellulose recovery from oil palm empty fruit bunch (EFB) fibers and production of sugars from the fibers. Carbohydr Polym 155:491–497
Patel M, Naik SN (2004) Gamma-oryzanol from rice bran oil–a review. J Sci Ind Res 63:569–578
Peanparkdee M, Iwamoto S (2019) Bioactive compounds from by-products of rice cultivation and rice processing: extraction and application in the food and pharmaceutical industries. Trends Food Sci Technol 86:109–117

Pintha K, Yodkeeree S, Limtrakul P (2015) Proanthocyanidin in red rice inhibits MDA-MB-231 breast cancer cell invasion via the expression control of invasive proteins. Biol Pharm Bull 38:571–581

Pintha K, Yodkeeree S, Pitchakarn P et al (2014) Anti-invasive activity against cancer cells of phytochemicals in red jasmine rice (*Oryza sativa* L.). Asian Pac J Cancer Prev 15:4601–4607

Prasad KN, Bondy SC (2019) Dietary fibers and their fermented short-chain fatty acids in prevention of human diseases. Bioact Carbohydr Diet Fiber 17:100170

Price RB, Shungu DC, Mao X et al (2009) Amino acid neurotransmitters assessed by proton magnetic resonance spectroscopy: relationship to treatment resistance in major depressive disorder. Biol Psychiatry 65:792–800

Pro B, Dang NH (2004) CD26/dipeptidyl peptidase IV and its role in cancer. Histol Histopathol 19:1345–1351

Rafe A, Sadeghian A, Hoseini-Yazdi SZ (2017) Physicochemical, functional, and nutritional characteristics of stabilized rice bran form tarom cultivar. Food Sci Nutr 5:407–414

Rao YPC, Pavan Kumar P, Lokesh BR et al (2016a) Molecular mechanisms for the modulation of selected inflammatory markers by dietary rice bran oil in rats fed partially hydrogenated vegetable fat. Lipids 51:451–467

Rao YPC, Sugasini D, Lokesh BR (2016b) Dietary gamma oryzanol plays a significant role in the anti-inflammatory activity of rice bran oil by decreasing pro-inflammatory mediators secreted by peritoneal macrophages of rats. Biochem Biophys Res Commun 479:747–752

Rastija V, Medić-Šarić M (2009) QSAR study of antioxidant activity of wine polyphenols. Eur J Med Chem 44:400–408

Ravindran V, Ravindran G, Sivalogan S (1994) Total and phytate phosphorus contents of various foods and feedstuffs of plant origin. Food Chem 50:133–136

Reiter E, Jiang Q, Christen S (2007) Anti-inflammatory properties of alpha- and gamma-tocopherol. Mol Asp Med 28:668–691

Renuka Devi R, Arumughan C (2007) Antiradical efcacy of phytochemical extracts from defatted rice bran. Food Chem Toxicol 45:2014–2021

Romeo B, Choucha W, Fossati P et al (2017) Meta-analysis of central and peripheral γ-aminobutyric acid levels in patients with unipolar and bipolar depression. J Psychiatry Neurosci 42:160228

Roohinejad S, Omidizadeh A, Mirhosseini H et al (2010) Effect of pre-germination time of brown rice on serum cholesterol levels of hypercholesterolaemic rats. J Sci Food Agric 90:245–251

Roohinejad S, Omidizadeh A, Mirhosseini H et al (2011) Effect of pre-germination time on amino acid profile and gamma amino butyric acid (GABA) contents in different varieties of Malaysian brown rice. Int J Food Prop 14:1386–1399

Rosenblat M, Volkova N, Borochov-Neori H et al (2015) Anti-atherogenic properties of date vs. pomegranate polyphenols: the benefits of the combination. Food Funct 6:1496–1509

Ryan EP, Heuberger AL, Weir TL et al (2011) Rice bran fermented with Saccharomyces boulardii generates novel metabolite profles with bioactivity. J Agric Food Chem 59:1862–1870

Saenjum C, Chaiyasut C, Chansakaow S et al (2012) Antioxidant and anti-inflammatory activities of gamma-oryzanol rich extracts from Thai purple rice bran. J Med Plant Res 6:1070–1077

Saija A, Tomaino A, Lo Cascio R et al (1999) Ferulic and caffeic acids as potential protective agents against photooxidative skin damage. J Sci Food Agric 79:476–480

Sakai S, Murata T, Tsubosaka Y et al (2012) γ-oryzanol reduces adhesion molecule expression in vascular endothelial cells via suppression of nuclear factor-κB activation. J Agric Food Chem 60:3367–3372

Schiraldi A, Fessas D (2012) Chapter 14—The role of water in dough formation and bread quality. In: Breadmaking: improving quality. Woodhead Publishing Series in Food Science, Technology and Nutrition, pp 352–369

Seesom C, Jumepaeng T, Luthria DL et al (2018) Comparison of extraction solvents and techniques used for the assay of free and bound phenolic acids from rice samples. J Food Health Bioenviron Sci 11:28–37

Sen CK, Khanna S, Rink C et al (2007) Tocotrienols: the emerging face of natural vitamin E. Vitam Horm 76:203–261

Serafim TL, Carvalho FS, Marques MPM et al (2011) Lipophilic caffeic and ferulic acid derivatives presenting cytotoxicity against human breast cancer cells. Chem Res Toxicol 24:763–774

Silva EO, Bracarense APFRL (2016) Phytic acid: from antinutritional to multiple protection of organic systems. J Food Sci 81:R135–R1362

Silveira ALM, Ferreira AVM, de Oliveira MC et al (2017) Preventive rather than therapeutic treatment with high fiber diet attenuates clinical and inflammatory markers of acute and chronic DSS-induced colitis in mice. Eur J Nutr 56:179–191

Singh A, Kukreti R, Saso L et al (2019) Oxidative stress: a key modulator in neurodegenerative diseases. Molecules 24:1583

Slavin J (2013) Fiber and prebiotics: mechanisms and health benefits. Nutrients 5:1417–1435

Soradech S, Reungpatthanaphong P, Tangsatirapakdee S et al (2016) Radical scavenging, antioxidant and melanogenesis stimulating activities of different species of rice (*Oryza sativa* L.) extracts for hair treatment formulation. Thai J Pharm Sci 40:92–95

Sut A, Pytel M, Zadrozny M et al (2019) Polyphenol-rich diet is associated with decreased level of inflammatory biomarkers in breast cancer patients. Rocz Panstw Zakl Hig 70:177–184

Tan BL, Norhaizan ME, Suhaniza HJ et al (2013) Antioxidant properties and antiproliferative effect of brewers' rice extract (*temukut*) on selected cancer cell lines. Int Food Res J 20:2117–2124

Tesfaye S, Wu J (2018) Diabetic neuropathy. In: Veves A, Giurini J, Guzman R (eds) The diabetic foot. Contemporary diabetes. Humana, Cham, pp 31–46

Tian J, Lu Y, Zhang H et al (2004) γ-aminobutyric acid inhibits T cell autoimmunity and the development of inflammatory responses in a mouse type 1 diabetes model. J Immunol 173:5298–5304

Tsai C-C, Chiu T-H, Ho C-Y et al (2013) Effects of anti-hypertension and intestinal microflora of spontaneously hypertensive rats fed γ-aminobutyric acid-enriched Chingshey purple sweet potato fermented milk by lactic acid bacteria. Afr J Microbiol Res 7:932–940

Tuura RLO, Baumann CR, Baumann-Vogel H (2018) Beyond dopamine: GABA, glutamate, and the axial symptoms of Parkinson disease. Front Neurol 9:806

Vogt SL, Pena-Diaz J, Finlay BB (2015) Chemical communication in the gut: effects of microbiota-generated metabolites on gastrointestinal bacterial pathogens. Anaerobe 34:106–115

Wanapat M, Sommart K, Saardrak K (1996) Cottonseed meal supplementation of dairy cattle fed rice straw. Livest Res Rural Dev 8:23

Wang Y-J, Maina NH, Ekholm P et al (2017) Retardation of oxidation by residual phytate in purified cereal β-glucans. Food Hydrocoll 66:161–167

Worasuwannarak N, Sonobe T, Tanthapanichakoon W (2007) Pyrolysis behaviors of rice straw, rice husk, and corncob by TG–MS technique. J Anal Appl Pyrolysis 78:265–271

Xia X, Ling W, Ma J et al (2006) An anthocyanin-rich extract from black rice enhances atherosclerotic plaque stabilization in apolipoprotein E-deficient mice. J Nutr 136:2220–2225

Xu J-G, Hu Q-P (2014) Changes in γ-aminobutyric acid content and related enzyme activities in Jindou 25 soybean (*Glycine max* L.) seeds during germination. LWT Food Sci Technol 55:341–346

Xu Z, Godber JS (1999) Purification and identification of components of gamma-oryzanol in rice bran oil. J Agric Food Chem 47:2724–2728

Yamakoshi J, Fukuda S, Satoh T et al (2007) Antihypertensive and natriuretic effects of less-sodium soy sauce containing γ-aminobutyric acid in spontaneously hypertensive rats. Biosci Biotechnol Biochem 71:165–173

Yang N-C, Jhou K-Y, Tseng C-Y (2012) Antihypertensive effect of mulberry leaf aqueous extract containing γ-aminobutyric acid in spontaneously hypertensive rats. Food Chem 132:1796–1801

Yatin SM, Varadarajan S, Butterfield DA (2000) Vitamin E prevents Alzheimer's amyloid-beta-peptide (1-42)-induced neuronal protein oxidation and reactive oxygen species production. J Alzheimers Dis 2:123–131

Yin Y, Cheng C, Fang W (2018) Effects of the inhibitor of glutamate decarboxylase on the development and GABA accumulation in germinating fava beans under hypoxia-NaCl stress. RSC Adv 8:20456–20461

Yu S, Nehus ZT, Badger TM et al (2007) Quantifcation of vitamin E and gamma-oryzanol components in rice germ and bran. J Agric Food Chem 55:7308–7313

Zhong S, Li L, Shen X et al (2019) An update on lipid oxidation and inflammation in cardiovascular diseases. Free Radic Biol Med 144:266–278

Zieliński H, Kozłowska H (2000) Antioxidant activity and total phenolics in selected cereal grains and their different morphological fractions. J Agric Food Chem 48:2008–2016

Zubair M, Anwar F, Shahid SA (2012) Effect of extraction solvents on phenolics and antioxidant activity of selected varieties of Pakistani rice (*Oryza sativa*). Int J Agric Biol 14:935–940

Chapter 5
Potential Health Benefits of Rice By-products

Abstract Rice by-products such as rice bran, straw, husk, and germ, are produced during the rice milling process. It has been considered as rich sources of bioactive components, for instance, γ-oryzanol, vitamin E, flavonoids, and phenolic compounds. These bioactive constituents are drawing interest due to the health benefits including antiobesity properties, hypoglycemic activity, chemopreventive effect, and cholesterol-lowering activity. In addition, rice by-products were shown to delay brain aging, decrease the risk of osteoporosis, and ameliorate arthritis. The beneficial effects of rice by-products were suggested to be partly mediated by their antioxidant effect of bioactive compounds. This chapter presents the biological mechanisms of rice by-products in the prevention of metabolic ailments. Collectively, a better understanding of the mechanism of action involved in rice by-products in metabolic ailments would provide a useful mean for the prevention of chronic diseases.

Keywords Cancer · Dementia · Diabetes · Hypercholesterolemic · Obesity · Osteoporosis

5.1 Antiobesity Activity

Obesity is a serious public health issue (Expert Panel Members et al. 2013; The GBD 2015 Obesity Collaborators 2017) characterized by a high body mass index (BMI) and more specifically excess abdominal fat. Obesity has been recognized as a major risk factor in the development of chronic diseases including diabetes, CVD, and cancer (Tan et al. 2018). It has been considered as a chronic low-grade inflammatory condition mediated by immune cells through the infiltration of adipose tissues, coupled with metabolic stress when oversupplied with lipids and glucose in adipocytes (Hotamisligil 2006; Oakes et al. 2013). Increased availability of inexpensive calorie-dense foods and sedentary lifestyles create an "obesogenic environment", which contributes to the obesity epidemic (Chaput et al. 2011; Yousefian et al. 2011; Ferdinand et al. 2012). Interaction of nutrients, especially dietary macronutrients, for instance, refined carbohydrates and saturated fats, and the individual heritability of obesity susceptibility genes, are associated with weight gain and thus may lead to

B. L. Tan, M. E. Norhaizan, *Rice By-products: Phytochemicals and Food Products Application*, https://doi.org/10.1007/978-3-030-46153-9_5

obesity (Spadaro et al. 2015; Bouchard 2008). Chronic intakes of high carbohydrate-high fat diet induce stress with the reduction of serum adiponectin, increased serum leptin, and abnormal adipokine secretion (Boonloh et al. 2015). Inflammatory cytokines are found in the fat cells, in which they are involved in fat metabolism and linked to all indices of obesity, especially abdominal obesity (de Heredia et al. 2012). Changes in fatty acid and oxidative stress levels, adipocyte function, and hypothalamic pituitary adrenal (HPA) axis and leptin dysfunction have been identified to play a crucial role in obesity-related inflammation (de Heredia et al. 2012). It has been demonstrated that physical activity and dietary food intake are the first choices to attenuate metabolic syndrome (Davi et al. 2010; Abete et al. 2011). Alternative treatments such as bariatric surgery and medication can also enhance weight loss and thereby maintaining body weight (Abdollahi and Afshar-Imani 2003). Nevertheless, the application of alternative therapies, for instance, the substitution of medication with changes in the diet in obese individuals is regarded as the potential approach in the prevention of obesity complications (Abete et al. 2011).

Dietary fiber has been evaluated for the first time by Keys et al. (1960). The data showed that dietary fiber was positively linked to long-term weight control as well as decreased the expression of IL-6 and C-reactive protein (CRP) (Johansson-Persson et al. 2014). In a randomized trial involving 105 overweight and obese adults demonstrated that intakes of rice bran and rice husk for 12 weeks reduced high-sensitivity (hs)-CRP and IL-6 levels (Edrisi et al. 2018) (Table 5.1). A few studies have also reported similar findings, in which dietary fiber has anti-inflammatory activity. Data from the human population-based study showed an inverse relationship between dietary fiber and hs-CRP in mildly hypercholesterolemic subjects (Johansson-Persson et al. 2014), indicating that dietary fiber may have a protective effect against hs-CRP. The mechanisms that implicate the amelioration of inflammation and dietary fiber require further elucidation; however, King (2005) has revealed that dietary fiber reduced inflammation by declining lipid oxidation. Importantly, dietary fiber also affects the production of SCFA in the intestine by the gut microbiota (Galisteo et al. 2008). Research evidence revealed that dietary

Table 5.1 The effect of rice by-products on obesity in animal and human studies

Animal and human studies	Rice by-products	Findings	References
Obese Zucker rats	Rice bran enzymatic extract	Enhanced a partial restoration of adiponectin levels	Justo et al. (2013)
Mice fed with a high-fat diet	40% stabilized rice bran	Decreased adipose tissue weight, body weight gain, and leptin levels	Kim et al. (2014)
Overweight and obese adults	Rice bran and rice husk	Reduced hs-CRP and IL-6 levels	Edrisi et al. (2018)
Obese rats	GABA enriched rice bran	Stimulates propionate and butyrate production by activating associated synthesizing enzymes, Anaerostipes sp., and Anaerostipes	Si et al. (2018)

GABA γ-aminobutyric acid, *hs-CRP* high-sensitivity C-reactive protein, *IL-6* interleukin-6

fibers are of benefit in alleviating IL-6 levels in adults (Grooms et al. 2013). Individuals with obesity or who are overweight showed an elevation of IL-6 serum level. Indeed, IL-6 is released from visceral adipose tissue into the portal blood circulation in obese individuals (Faam et al. 2014).

Rice bran enzymatic extract is produced by an enzymatic process that improves the antioxidant and protein solubility and preserves the functional properties of rice bran (Parrado et al. 2003, 2006). Feeding obese Zucker rats with rice bran enzymatic extract improved imbalanced production of $NO_{(x)}$ and enhanced a partial restoration of adiponectin levels, implying the potential role of enzymatic extract on systemic inflammation (Justo et al. 2013). The primary components of rice bran enzymatic extract involved in the modulation of inflammation-associated obesity and hypoadiponectinemia could be due to the presence of γ-oryzanol and ferulic acid. The previous finding has corroborated this study and found that γ-oryzanol restored adiponectin levels in obese mice (Nagasaka et al. 2011). Moreover, Kim et al. (2014) found that 40% of stabilized rice bran decreased adipose tissue weight, body weight gain, and leptin levels in mice fed with a high-fat diet. Reduced leptin levels may stimulate a reduction of hypertension (Lane and Vesely 2013). Notably, leptin deficiency is related to the early onset of obesity, suggesting that ratio of leptin to insulin is required in the homeostatic balance of glucose and fat metabolism (Wabitsch et al. 2015). Leptin, a white adipose tissue-derived hormone, has been demonstrated to enhance ROS accumulation in endothelial cells (Maingrette and Renier 2003; Yamagishi et al. 2001). Chronic administration of a high-fat diet may trigger the resistance of leptin (Knight et al. 2010). An animal study showed that GABA enriched rice bran stimulates propionate and butyrate production by activating associated synthesizing enzymes, Anaerostipes sp., and Anaerostipes and thereby altering SCFA and further increased the glucagon-like peptide-1 and circulatory levels of leptin (Si et al. 2018). Overall, the evidence demonstrated that rice by-products have the potential to promote weight loss (Fig. 5.1). Further studies are required to evaluate the precise molecular mechanism to better delineate the role of rice by-products on weight management.

5.2 Chemopreventive Effect

Cancer has become a significant health problem affecting women and men globally. Oxidative stress is implicated in cancer development (Fiaschi and Chiarugi 2012; Tan et al. 2018). Increased ROS levels contribute to the susceptibility to mutagenic agents or elevation of mutation rates, thereby leads to DNA damage in the initial stage of carcinogenesis (Hanahan and Weinberg 2011). Elevation of ROS promotes tumor proliferation through ligand-independent transactivation of receptor tyrosine kinase (Fiaschi and Chiarugi 2012), and subsequently leads to invasion and metastasis of cancer cells (Gupta et al. 2012). ROS can stimulate the stabilization of hypoxia-inducible factor 1, a molecule for vascular endothelial growth factor, which modulates the activity of tumor angiogenesis (Semenza 2010). Nearly 90% of all

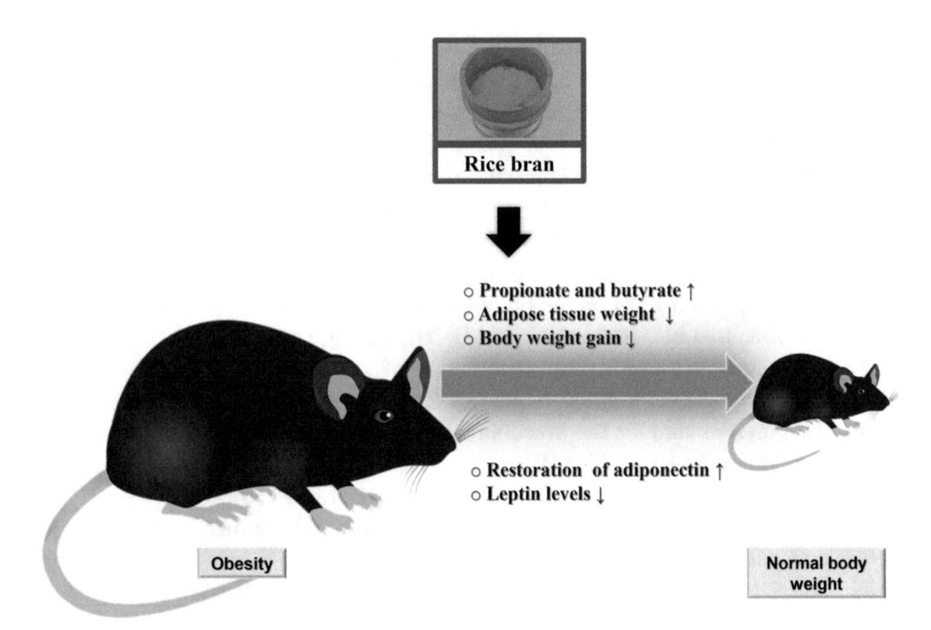

Fig. 5.1 The interaction of rice bran in relation to weight loss

cancers directly or indirectly linked to an individual's dietary habits and lifestyles, while the rest is contributed by genetic factors (Anand et al. 2008). Research evidence has shown that oxidative stress and cancer can be modulated by nutrients-rich in antioxidants and phytochemicals (Tan et al. 2018). Antioxidants facilitate the breaking and termination of an oxidative chain reaction through inhibiting the formation of free radical intermediates (Gholamian-Dehkordi et al. 2017). A unique complex of bioactive components protects against oxidative stress that contributes to inflammation (Tan et al. 2015a, b; Tan and Norhaizan 2017). Figure 5.2 illustrates the effects of oxidative stress and interactions of rice by-products and cancer.

Rice by-products are valuable source of phytochemicals for the prevention of cancer both *in vivo* and *in vitro* studies. Relative portions of bioactive compounds in rice bran have been demonstrated to suppress the proliferation of several cancer cells such as colorectal cancer (Forster et al. 2013). Tables 5.2 and 5.3 summarize the chemopreventive effects of rice by-products in both *in vivo* and *in vitro* studies. Bioactive constituents such as flavonoids (Chang et al. 2018), tricin (Malvicini et al. 2018), and rice bran phytic acid (Tan and Norhaizan 2017) have been demonstrated to inhibit the growth of colorectal cancer cells (Table 5.2); however, the effects are varied among rice varieties (Okonogi et al. 2018). Chen et al. (2012) utilized several cancer cells to evaluate the cell-suppressing activity on red rice bran extract and found that red bran possesses strong suppressive effects on stomach, cervical, and leukemia cancers. Such finding highlights the role of unique complexes of bioactive constituents in rice by-products. A study on the effect of proanthocyanidin-rich fraction obtained from red rice bran and germ extract on hepatocellular carcinoma (HepG2) cells was conducted by Upanan et al. (2019).

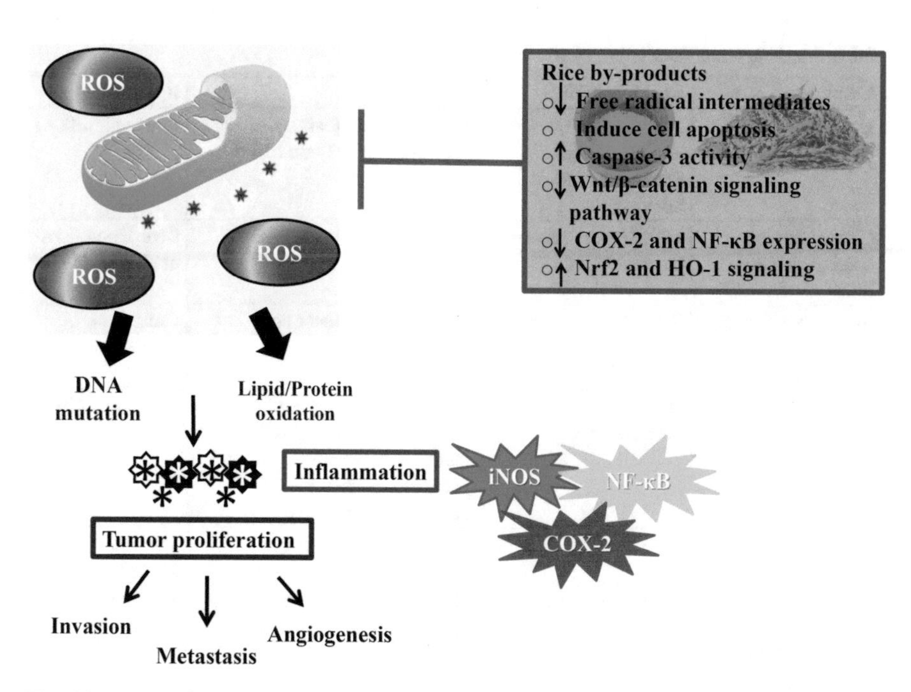

Fig. 5.2 The effect of oxidative stress and the interaction of rice by-products on cancer. *COX-2* cyclooxygenase-2, *DNA* deoxyribonucleic acid, *HO-1* heme oxygenase-1, *iNOS* inducible nitric oxide synthase, *NF-κB* nuclear factor-kappa B, *Nrf2* NF-E2-related factor 2, *ROS* reactive oxygen species

They found that proanthocyanidin-rich fraction derived from rice bran and germ extract could induce cell apoptosis and suppress cell proliferation in HepG2 cells by decreasing survivin protein levels and increasing apoptotic protein levels including cleaved caspase-3 and cleaved caspase-8 activities (Upanan et al. 2019). Besides rice bran and germ, rice straw extract has shown a promising inhibitory effect against breast carcinoma (MCF-7), prostate (PC-3), and liver cancer cell lines (Meselhy et al. 2018). This finding implied that rice straw extract is capable to suppress tumor proliferation predominantly due to the antioxidant activity and phenolic compounds. Research evidence indicates that phenolic compounds are of benefit in the induction of G_0/G_1 phase and upregulation of caspase-3 activity (Wang et al. 2016). In addition, a few studies have indicated that certain phytochemicals in rice bran would facilitate the suppression of cell proliferation via modulation of Wnt/β-catenin, predominantly via a reduction in β-catenin expression. Beta-catenin is a protein that serves as a mediator in the Wnt signaling, which is known to contribute to cell proliferation. Aberrant stimulation of β-catenin has been associated with the development of cancer (Polakis 2012). The essential role of c-Myc in the modulation of cell growth by Wnt pathway and β-catenin is well-recognized (He et al. 1998), while cyclin D1 is another cell cycle protein targeted by β-catenin (Tetsu and McCormick 1999) that is usually overexpressed in tumor tissues. The previous study showed that methanol extract of Riceberry bran reduced cyclin D1 expression

Table 5.2 The anticancer effect of rice by-products *in vitro*

Cell lines	Rice by-products	Findings	References
Human lymphoma (Jurkat) cells	Momilactone B, allelochemical extracted from rice husk	Induced apoptosis by stimulating caspase activity	Lee et al. (2008)
Caco-2 colonic carcinoma cell lines	Methanol extract of Riceberry bran	Reduced cyclin D1 expression	Leardkamolkarn et al. (2011)
Stomach, cervical, and leukemia cancer cells	Red rice bran extract	Suppressed cancer cell proliferation	Chen et al. (2012)
Human colorectal cancer (HT-29) cells	Water extract of brewers' rice	Upregulated *APC* and *CK1* and downregulated *GSK3β* and *LRP6* mRNA levels and activated caspase-3 and caspase-8 activities	Tan et al. (2015a)
Colorectal cancer cells	Rice bran phytic acid	Suppressed cancer cell proliferation	Tan and Norhaizan (2017)
Breast carcinoma (MCF-7), prostate (PC-3), and liver cancer cells	Rice straw extract	Inhibited cancer cell proliferation	Meselhy et al. (2018)
Hepatocellular carcinoma (HepG2) cells	Proanthocyanidin-rich fraction derived from rice bran and germ extract	Induced cell apoptosis and suppress cell proliferation by decreasing survivin protein levels and increasing apoptotic proteins such as cleaved caspase-3 and -8 activities	Upanan et al. (2019)

APC adenomatous polyposis coli, *CK1* casein kinase 1, *GSK3β* glycogen synthase kinase 3β, *LRP6* low-density lipoprotein receptor-related protein 6

in Caco-2 colonic carcinoma cell lines (Leardkamolkarn et al. 2011). Reduced cyclin D1 in colon carcinoma cells has been reported due to the tocotrienol or vitamin E content (Gysin et al. 2002). Vitamin E derivatives including δ- and γ-tocotrienols were found to decrease the nuclear localization and expression of β-catenin in colon cancer cell lines, thereby leading to a reduction of c-myc and cyclin D1 levels (Xu et al. 2012; Zhang et al. 2011). This finding suggests that rice bran phytochemicals may disrupt the Wnt/β-catenin pathways via a reduction of β-catenin expression. Suppression of colon cancer through inhibition of Wnt/β-catenin signaling was also described in brewers' rice. Water extract of brewers' rice was shown to upregulate adenomatous polyposis coli (*APC*) and casein kinase 1 (*CK1*) and downregulate the glycogen synthase kinase 3β (*GSK3β*) and low-density lipoprotein receptor-related protein 6 (*LRP6*) mRNA levels (Tan et al. 2015a). APC, GSK3β, and CK1 are destruction complex involved in the phosphorylation of β-catenin, while LRP6 is a critical coreceptor in Wnt signaling. Downregulation of *GSK3β* expression observed in this study implies that other mechanisms are

Table 5.3 The anticancer effect of rice by-products *in vivo*

Animals	Rice by-products	Findings	References
Rats	Rice germ	Inhibited colon adenocarcinoma	Kawabata et al. (1999)
Rats	Rice germ	Reduced ACF/colon	Mori et al. (1999)
Fisher 344 rats	Defatted rice germ	Reduced the incidence of tongue carcinoma	Mori et al. (1999)
Male F344 rats	Methanolic extract of rice husk	Reduced ACF formation	Kim et al. (2007)
Rats	p-methoxycinnamic acid, rice bran phenolic acid	Reduced the formation of aberrant crypt foci and reversed the reduction of antioxidant enzymes caused by dimethylhydrazine-induced colon cancer	Sivagami et al. (2012)
Male Sprague-Dawley rats	Brewers' rice	Reduced colon tumor multiplicity	Tan et al. (2014)
Male Sprague-Dawley rats	Brewers' rice	Enhanced antioxidant enzymes such as NO, MDA, and SOD and enhanced Nrf2 signaling pathway	Tan et al. (2015b)
Hamsters	5% or 10% fermented brown rice and rice bran	Inhibited BOP-induced pancreatic cancer	Kuno et al. (2015)
Swiss albino mice	MGN-3/Biobran	Enhanced radiation therapy by activating caspase-3 and Bax/Bcl-2 ratio, downregulating Bcl-2 expression, and upregulating the protein and relative gene expression of Bax and p53 in tumor cells	El-Din et al. (2019)
Ehrlich solid carcinoma-bearing mice	Rice straw with a low dosage of gamma radiation	Synergistically inhibited the murine Ehrlich solid carcinoma via stimulation of caspase-3 activity and Bax expression	Meselhy et al. (2019)

ACF aberrant crypt foci, *BOP* *N*-Nitrosobis(2-oxopropyl)amine, *MDA* malondialdehyde, *MGN-3/Biobran* arabinoxylan rice bran, *NO* nitric oxide, *Nrf2* NF-E2-related factor 2, *SOD* superoxide dismutase

involved in the water extract of brewers' rice and are likely mediated by NF-κB (Billadeau 2007; Ougolkov and Billadeau 2006). Treatment water extract of brewers' rice also induces apoptosis in colon cancer (HT-29) cell lines by stimulating caspase-8 and caspase-3 activities (Tan et al. 2015a). Besides rice bran and brewers' rice, Momilactone B, an allelochemical extracted from rice husk, triggers apoptosis in human lymphoma (Jurkat) cells by stimulating caspase activity (Lee et al. 2008).

The chemopreventive potential of bioactive constituents in rice by-products was not only observed in *in vitro* study, they have also been demonstrated *in vivo* models. Intakes of 30% fibers from bran reduced adenoma burden in mice, which is about 1.2 g/day/kg rice bran fiber for human consumption (Norris et al. 2015). Dietary rice bran is believed to have chemopreventive activity by activating caspase-3 expression, inhibiting COX-2, and modulating NF-κB signaling and iNOS expression (Phutthaphadoong et al. 2010; El-Din et al. 2016; Tan and Norhaizan 2017). Several studies have indicated that rice by-products can reduce chemical-induced carcinogenesis (Esa et al. 2013) (Table 5.3). For instance, a diet containing 5% or 10% rice bran and fermented brown rice was shown to inhibit *N*-Nitrosobis(2-oxopropyl)amine (BOP)-induced pancreatic cancer in hamsters (Kuno et al. 2015). In support of this, feeding with p-methoxycinnamic acid, a rice bran phenolic acid, reduces the formation of aberrant crypt foci (ACF) in dimethylhydrazine-induced colon cancer rats (Sivagami et al. 2012). This study further revealed that p-methoxycinnamic acid reverses the reduction of antioxidant enzymes caused by the carcinogen (Sivagami et al. 2012). Such findings suggest that p-methoxycinnamic acid treatment confers protection against colon cancer has been attributed to the induction of apoptosis (Gunasekaran et al. 2015). In addition, feeding rats with rice germ (5 weeks) significantly inhibited colon adenocarcinoma in azoxymethane (AOM)-induced rats (Kawabata et al. 1999). Likewise, Mori et al. (1999) found that dietary administration of 2.5% rice germ significantly reduced ACF/colon in rats compared to the control group. While in tongue cancer, Mori et al. (1999) showed that defatted rice germ can reduce the incidence of tongue carcinoma in Fisher 344 rats. The suppressive effect of colon preneoplastic ACF was also demonstrated in rice husk. The data showed that administration of methanolic extract of rice husk for 40 weeks reduced ACF formation in 1,2-dimethylhydrazine-induced male F344 rats (Kim et al. 2007). Chronic inflammation and infection activate the inflammatory-associated genes such as *iNOS* (van der Woude et al. 2004) and NF-κB levels (Pikarsky et al. 2004). Feeding rats with brewers' rice for 20 weeks, a rice by-product consisting of rice germ, rice bran, and broken rice, could reduce the colon tumor multiplicity (Tan et al. 2014). In addition, Tan et al. (2015b) further revealed that brewers' rice decreased oxidative stress in AOM-induced rats colon cancer. The data showed that dietary administration of brewers' rice enhanced antioxidant enzymes such as NO, malondialdehyde (MDA), and superoxide dismutase (SOD) (Tan et al. 2015b). The study also reported that brewers' rice enhanced NF-E2-related factor 2 (Nrf2) signaling pathway through heme oxygenase-1 (*HO*-1) and *Nrf2* mRNA levels (Tan et al. 2015b), suggesting that rice by-products inhibit oxidative stress may be modulated partly through additive/synergistic effects of bioactive compounds.

Besides decreasing chemical-induced carcinogenesis, rice-by-products can also improve the sensitivity of cancer cells to chemotherapy drugs. In a study focused on breast cancer outcomes, El-Din et al. (2019) evaluated the ability of arabinoxylan rice bran (MGN-3/Biobran) to enhance the anticancer effects in fractionated X-ray irradiation of Ehrlich solid tumor-bearing mice. The data showed that administration of Biobran (40 mg/kg/day, intraperitoneal injections; starting from day 11 (post-tumor cell inoculation) until day 30) + ionizing radiation promotes caspase-3 and Bax/Bcl-2 ratio, downregulated Bcl-2 expression, and upregulated the protein and relative gene expression of Bax and p53 in tumor cells (El-Din et al. 2019). This finding indicates that Biobran increases radiation-induced tumor regression by modulating apoptosis and minimizing toxicity-associated radiation therapy (El-Din et al. 2019). In support of this, the combination of rice straw with a low dosage of gamma radiation synergistically inhibits the murine Ehrlich solid carcinoma via stimulation of caspase-3 activity and Bax expression in Ehrlich solid carcinoma-bearing mice (Meselhy et al. 2019). Such finding highlights that low dose ionizing radiation could suppress the cancer cells proliferation via a few mechanisms such as increased cellular anticancer immunity, triggers hormesis in the immune system by activating T cells growth, promote antibody-dependent cellular cytotoxicity response, and surge antibody secretion in tumor-bearing mice, which is well-correlated with tumor regression (Kojima et al. 2004). Furthermore, bioactive constituents in rice by-products have also been reported to facilitate cancer chemoprevention via an immune response. The immune response serves either through specific inhibition of tumor cells or blocking viral infections associated with virus-induced tumors (Vesely et al. 2011).

Rice bran may increase intestinal SCFA production and decrease intestinal pH via modulation of gut microbiota (Sheflin et al. 2015). Interactions between microbial and host metabolism during colonic fermentation lead to the production of secondary metabolites (Schmidt et al. 2014). Modulating the substrates available to gut microbiota can affect the amount and type of microbial metabolites produced, and thus may contribute to metabolic disorders including cancer (Hullar et al. 2014). Based on the findings of rice bran consumption in the clinical trials, changes in microbial and host metabolism would be expected to modulate the stool metabolite profile. This study showed that heat-stabilized rice bran consumption (30 g/day for 4 weeks) favorably mediated the stool metabolome in colorectal cancer survivors (Brown et al. 2017) (Table 5.4). Notably, the suppressive effects of bile acid metabolites (secondary and primary) observed in colorectal cancer survivors are similar to healthy adults (Sheflin et al. 2015). In general, primary bile acids are synthesized by the human liver and converted to secondary bile acids by the microbiota (Sayin et al. 2013). Secondary bile acids have been linked to the ROS/RNS production and oxidative DNA damage (Payne et al. 2008). Similarly, Sheflin et al. (2017) observed that intakes of heat-stabilized rice bran (30 g/day for 28 days) altered gut microbial composition and increased gut bacterial diversity in colorectal cancer survivors, implied the ability of heat-stabilized rice bran to improve gut health. The improvement in these indices could be due to the phytochemicals and dietary fibers (Tan et al. 2013). Further, data from a randomized controlled pilot clinical trial involving 29 overweight/obese volunteers (with a history of colorectal cancer) consumed heat-stabilized rice bran for 28 days altered gut microbial composition and increased gut bacterial diversity (Sheflin et al. 2017). This finding implies that consumption of high-fiber foods such as heat-stabilized rice bran has been linked to a reduced risk of colorectal cancer (Egeberg et al. 2010; Fung et al. 2010), which may lead to the changes in gut microbiota activities. Besides colorectal cancer, oral intake of hydrolyzed rice bran can ameliorate the diarrheal adverse outcome of chemotherapy in patients with cervical cancer (Itoh et al. 2015). Bang et al. (2010) further demonstrated that intakes of MGN-3/Biobran significantly increased the two-year survival rate and reduced the recurrence in hepatocellular carcinoma patients. Similarly, multiple myeloma patients supplemented with MGN-3/Biobran has also been

Table 5.4 The anticancer effect of rice by-products in human studies

Subjects	Rice by-products	Findings	References
Leukemia, breast, multiple myeloma, and prostate cancer patients	MGN-3/Biobran	Increased natural killer activity and enhanced different cytokines levels	Ghoneum and Brown (1999)
Hepatocellular carcinoma patients (stages I and II)	MGN-3/Biobran	Reduced recurrence and increased two-year survival rate	Bang et al. (2010)
Multiple myeloma patients	MGN-3/Biobran	Increased in natural killer activity, enhanced level of myeloid dendritic cells in peripheral blood, and augmented concentrations of T helper cell type 1-related cytokines	Cholujova et al. (2013)
Cervical cancer patients	Hydrolyzed rice bran	Attenuated diarrheal side effects of chemotherapy	Itoh et al. (2015)
Colorectal cancer survivors	Heat-stabilized rice bran (30 g/day for 4 weeks)	Modulated stool metabolome	Brown et al. (2017)
Colorectal cancer survivors	Heat-stabilized rice bran (30 g/day for 28 days)	Altered gut microbial composition and increased gut bacterial diversity	Sheflin et al. (2017)

MGN-3/Biobran arabinoxylan rice bran

reported to increase the concentrations of T helper cell type 1-related cytokines, enhance the levels of myeloid dendritic cells in peripheral blood, and increase the NK activity (Cholujova et al. 2013). NK cells are crucial in innate immunity for immunotherapy and cancer prevention. Inhibition of NK activity in cancer patients was linked to defective lymphokine production. Treatment with MGN-3/Biobran markedly increased NK activity in cancer patients including leukemia, breast, multiple myeloma, and prostate (Ghoneum and Brown 1999). It appears that MGN-3/Biobran augmented the NK cell cytotoxic function and enhanced the levels of different cytokines (Ghoneum and Brown 1999), suggesting that MGN-3/Biobran could enhance the overall immune activity. The beneficial effects of rice by-products in combating carcinogenesis have been suggested partly mediated via additive/synergistic effects of bioactive components (Jacobs and Tapsell 2007). The bioactive components in rice by-products were shown a better inhibitory effect in cancer perhaps even better than drugs, implied that whole food extract or whole food is vitally important in the management of cancer (Ricciardiello et al. 2011). In support of this, Tsuda et al. (2004) also reported that a synergistic/additive effect of some bioactive components. Although preclinical findings and clinical trials have shown a promising effect of rice by-products in combating cancer, further studies are crucial to elucidate the relationship between rice by-products and cancer in the long-term clinical trials.

5.3 Cholesterol-Lowering Activity

CVD is the disorder of the heart and blood vessels such as rheumatic heart disease, coronary heart disease, cerebrovascular disease, and other conditions. About 85% of all CVD deaths are due to strokes and heart attacks. Individuals at risk of CVD often showed a high lipid level (World Health Organization 2019a). According to the World Health Organization (2019a), nearly 17.9 million people are killed due to the CVD annually, which is contributed 31% of all deaths globally. It has been reported that increased vascular production of ROS, especially superoxide causes detrimental impacts in cell death (Ohara et al. 1993; Hink et al. 2001). RNS and ROS are tightly related to the disease process. Incomplete scavenging of RNS and ROS affects the mitochondrial lipid cardiolipin, enhances the release of mitochondrial cytochrome c and ultimately stimulates the intrinsic death signaling (Victor et al. 2009). Local RNS production may lead to vascular tissue injury. In this regard, RNS and ROS may participate as signaling molecules that mediate a broad spectrum of pathophysiological pathways. The pathogenesis of atherosclerosis is further linked to the proliferative process, immune response, and inflammation (Raggi et al. 2018). In particular, endothelial denuding injury could contribute to the platelet aggregation and enhance the release of platelet-derived growth factor, and thereby increase the proliferation of smooth muscle cells forming the nidus of the atherosclerotic plaque in the arterial intima, indicating the inflammatory response has implicated in the development of CVD (Libby et al. 2009; Sugamura and Keaney 2011).

CVD is being a dietary pattern as a major and modifiable exposure risk factor. Consumption of foods rich in bioactive constituents and dietary fibers can prevent or treat CVD, due to its potential to decrease hyperlipidemia, oxidative stress, and inflammation (Esa et al. 2013; Singh et al. 2019). Experimental and epidemiological studies have demonstrated the cardiometabolic protection of different rice bran extracts, RBO, rice bran, alone or in diet supplementation (Saji et al. 2019). The previous study revealed that rice bran exerts hypolipidemic activity by enhancing HDL-C and reducing TG and total cholesterol (TC) levels in primates, rabbits, and rodents (Cicero and Gaddi 2001) (Table 5.5). Recent studies suggest the beneficial effects of rice by-products in relation to CVD (Senaphan et al. 2018). Rice bran is a nutraceutical ingredient that affects dyslipidemia (Ijiri et al. 2015). The hypolipidemic activity of rice bran was associated with increased bile acid synthesis (Kritika and Virginia 2018) and HDL-C (Newman et al. 1992) via induction of CYP7A1 (Matheson et al. 1995) and thereby leading to higher cholesterol excretion (Ijiri et al. 2015). Similarly, data from the animal study revealed that supplementation of rice bran extract (10–50 g/kg) reduced TC, triacylglycerol, COX-2, and iNOS levels in C57BL/6 ApoE−/− mice (Perez-Ternero et al. 2017). This finding is further supported by Justo et al. (2016), who found that feeding C57BL/6 mice with rice bran extract reduced TC, triacylglycerol, IL-6, and IL-1β levels. Figure 5.3 summarizes the detrimental effects of oxidative stress in relation to CVD and the mechanisms of rice by-products involved in the modulation of CVD.

Table 5.5 The effects of rice by-products on lipid profile in *in vitro* and animal studies

Animal studies	Rice by-products	Findings	References
Cow	Roughage neutral detergent fiber from rice straw	Reduced β-hydroxymethylglutaryl-CoA reductase and activated β-hydroxymethylglutaryl-CoA synthase	Grummer and Carroll (1988)
Primates, rabbits, and rodents	Rice bran	Improved HDL-C and decreased TG and TC levels	Cicero and Gaddi (2001)
Dairy cows	Neutral detergent fiber from rice straw	Increased LDL-C levels but not affected HDL-C, TC, and TG levels	Kanjanapruthipong and Thaboot (2006)
C57BL/6 mice	Rice bran extract	Reduced TC, triacylglycerol, IL-6, and IL-1β levels	Justo et al. (2016)
C57BL/6 ApoE−/− mice	Rice bran extract	Reduced TC, triacylglycerol, COX-2, and iNOS levels	Perez-Ternero et al. (2017)
RAW264.7 cell lines	Rice bran δ-tocotrienol	Suppressed proinflammatory cytokines (IL-6, IL-1β, IFN-γ, and TNF-α) and LPS-stimulated NO and inhibited the phosphorylation of ERK1/2 and JNK	Shen et al. (2018)
Weaned piglets	HEBR	No effect on cholesterol, TG, LDL-C or HDL-C	Huang et al. (2019)
Hypercholesterolemic rats	Rice bran	Reduced LDL-C, TC, and atherogenic index	Nandi et al. (2019)

COX-2 cyclooxygenase-2, *CVD* cardiovascular disease, *ERK1/2* extracellular regulated protein kinases 1/2, *HDL-C* high-density lipoprotein cholesterol, *HEBR* hydrolysates produced by limited enzymatic broken rice, *IFN-γ* interferon gamma, *iNOS* inducible nitric oxide synthase, *IL-1β* interleukin-1beta, *IL-6* interleukin-6, *JNK* c-Jun N-terminal kinase, *LDL-C* low-density lipoprotein cholesterol, *LPS* lipopolysaccharides, *NO* nitric oxide, *TC* total cholesterol, *TG* triglycerides, *TNF-α* tumor necrosis factor alpha

Traditionally, vitamin E has been linked to antioxidant activities by upregulating antioxidant enzymes and scavenging ROS (Abraham et al. 2019). Vitamin E is believed to have hypolipidemic activity through degradation and ubiquitination of HMG-CoA reductase and inhibiting the process of sterol regulatory element-binding proteins (SREBPs) (Song and DeBose-Boyd 2006). Intriguingly, tocotrienols exert better suppression of triacylglycerol and serum cholesterols in Fischer 344 rats fed a western diet compared to α-tocopherols (Shibata et al. 2016). The hypocholesterolemic potential of vitamin E in RBO has been demonstrated in several animal models (Pearce et al. 1992; Esa et al. 2013). Feeding rats with tocotrienols (200–400 mg/kg body weight) increased HDL-C, SOD, glutathione reductase, glutathione peroxidase, and catalase levels (Siddiqui et al. 2013). In a study focused on inflammation outcomes, Shen et al. (2018) evaluated the mechanism and anti-inflammatory effect of rice bran δ-tocotrienol against lipopolysaccharides (LPS)

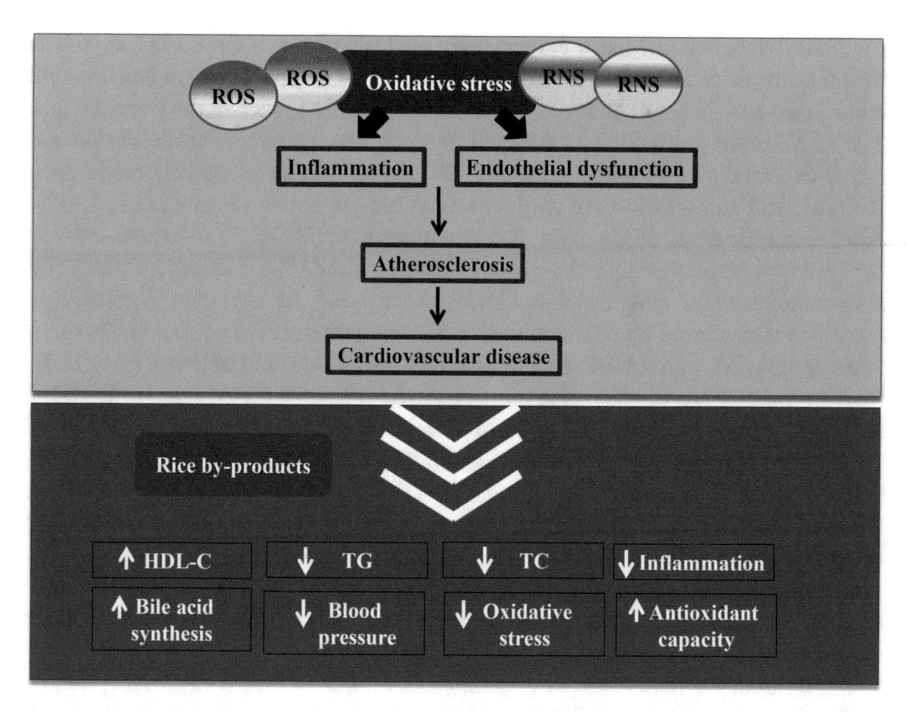

Fig. 5.3 Detrimental effects of oxidative stress in relation to cardiovascular disease (CVD) and the mechanisms of rice by-products involved in the modulation of CVD. *HDL-C* high-density lipoprotein cholesterol, *RNS* reactive nitrogen species, *ROS* reactive oxygen species, *TC* total cholesterol, *TG* triglycerides

activated proinflammatory mediator expressions in RAW264.7 cell lines. The data showed that rice bran δ-tocotrienol suppressed proinflammatory cytokines (IL-6, IL-1β, interferon-gamma (IFN-γ), and TNF-α) and LPS-stimulated NO and inhibited the phosphorylation of extracellular regulated protein kinases 1/2 (ERK1/2) and c-Jun N-terminal kinase (JNK). The study found that rice bran δ-tocotrienol inhibited inflammation through MAPK and PPAR pathways in LPS-activated macrophages (Shen et al. 2018). PPAR is a crucial transcriptional factor involved in the inflammation and lipid metabolism, in which its activity is tightly linked to the development of hyperlipidemia and chronic inflammation (Barish et al. 2006).

Total dietary fiber is a mixture of components present in plant-based foods that are not digested in the small intestine, in which these components contained plenty of lignins and non-starchy carbohydrates that are intrinsic to the plant and remain structurally intact such as resistant starch and cell wall components (Evans 2019). Consumption of a diet rich in dietary fiber decreasing hypercholesterolemia via elevation of the excretion of cholesterol through fecal routes due to the resistance to digestion by human gastrointestinal enzymes (Gallaher et al. 2002). A study by Nandi et al. (2019) evaluated the functional properties of defatted sesame husk, rice bran, and flaxseed on hypercholesterolemic rats. The study showed that feeding 3 g/100 g of total dietary fiber extracted from sesame husk, flaxseed, and rice bran for 28 days reduced low-density lipoprotein cholesterol (LDL-C) and TC, as well as the atherogenic index in hypercholesterolemic rats. Gamma-oryzanol has shown a better lipid-lowering effect due to the mechanism taking place in the liver and gut. The sterol moiety promotes cholesterol excretion by upregulating the basolateral sterol exporter ATP-binding cassette (ABC-A) or physicochemical interference with micellar solubilization of cholesterols in the gut lumen (Brauner et al. 2012). After absorption, the ferulic acid moiety suppresses HMG-CoA reductase-derived synthesis of cholesterol in the liver (Wang et al. 2015). Notably, a study by Kanjanapruthipong and Thaboot (2006) found that LDL-C was significantly increased after dietary treatment with neutral detergent fiber from rice straw in dairy cows ($p < 0.05$), while HDL-C, TC, and TG levels were not affected ($p > 0.05$). Increased acetyl-CoA in lactation from a cow fed with total mixed rations (TMR) of roughage neutral detergent fiber from rice straw may lead to the reduction of β-hydroxymethylglutaryl-CoA reductase and activation of β-hydroxymethylglutaryl-CoA synthase. The depression may inhibit LDL receptors synthesis and enhance cholesterol acyltransferase activity (Grummer and Carroll 1988). However, feeding weaned piglets with hydrolysates produced by limited enzymatic broken rice (HEBR) (30 g/kg for 21 days) did not significantly reduce the TG, cholesterol, and LDL-C or increase HDL-C levels compared to the control ($p > 0.05$) (Huang et al. 2019).

In addition, the data revealed that supplementation of rice bran extracts rich in acylated steryl glucosides (30–50 mg) reduced LDL-C in patients with obese and high LDL-C (Ito et al. 2015) (Table 5.6). It has been demonstrated that vitamin E, particularly tocotrienol-rich fraction, showed the most suppressive effect on inflammation and hypercholesterolemia compared to α-tocopheryl acetate and α-tocopherol (Ng and Ko 2012). In a study reported by Ajuluchukwu et al. (2007) and Baliarsingh et al. (2005) focusing on rice bran tocotrienols and hypercholesterolemia outcome, rice bran tocotrienols were shown to improve lipid profiles, suggesting its ability to treat and prevent hyperlipidemia. Such finding highlights the unique structure of tocotrienols with the presence of three double bonds in the side chain, and thus confers better anti-inflammatory and antioxidant activities compared to tocopherols. The differential effects between tocotrienols and tocopherols are more likely due to the unsaturated side chain of tocotrienols that allows more efficient penetration into the tissues and distribution in the cell membranes with saturated fatty layers including liver and brain (Suzuki et al. 1993). In a study by Erlinawati et al. (2017) focusing on the RBO supplementation (45 mL/day or 15 mL/day) and lipid profile in mild-moderate hypercholesterolemic males aged 19–55 years old, consumption of 45 mL/day RBO showed better lipid profiles than 15 mL/day. The data showed that group supplemented with 45 mL/day of RBO reduced TC by 14%, while 15 mL/day reduced TC by 7.8%, suggesting that consumption of 45 mL/day RBO led to better improvements in lipid profiles. However, the levels of HDL-C, LDL-C, and TG were not significantly different between both groups ($p > 0.05$). Such finding implies that plant sterols and γ-oryzanol in RBO exert an ability to remove cholesterol from bile salt micelles and thereby reducing cholesterol absorption in the intestine (Erlinawati et al. 2017). Likewise, the randomized double-blind controlled trial

Table 5.6 The effects of rice by-products on lipid profile in human studies

Subjects	Rice by-products	Findings	References
Patients with mildly hypercholesterolemia or type 2 diabetic patients with hyperlipidemia	Rice bran tocotrienols	Improved lipid profiles	Baliarsingh et al. (2005); Ajuluchukwu et al. (2007)
Obese and high LDL-C patients	Rice bran extract rich in acylated steryl glucosides	Reduced LDL-C levels	Ito et al. (2015)
Mild-to-moderate hypertensive patients	Sesame oil and RBO (20:80)	Increased HDL-C and reduced LDL-C and blood pressure	Devarajan et al. (2016)
Postmenopausal women with type 2 diabetes	RBO	Reduced TG, TC, LDL-C, non-HDL-C levels	Salar et al. (2016)
Children with abnormal lipid profile	Rice bran	No association was observed between rice bran and TC, HDL-C, LDL-C, or triglycerides	Borresen et al. (2017)
Mild-moderate hypercholesterolemic males	RBO (45 mL/day or 15 mL/day)	Supplementation of 45 mL/day RBO showed better lipid profiles than 15 mL/day	Erlinawati et al. (2017)
Hyperlipidemic subjects	RBO containing γ-oryzanol	Reduced LDL-C levels and increased antioxidant capacity	Bumrungpert et al. (2019)
CVD patients	Rice bran extracts, RBO, or rice bran	Improved cardiometabolic function	Saji et al. (2019)

CVD cardiovascular disease, *HDL-C* high-density lipoprotein cholesterol, *LDL-C* low-density lipoprotein cholesterol, *RBO* rice bran oil, *TC* total cholesterol, *TG* triglycerides

involving 59 hyperlipidemic subjects revealed that intakes of RBO containing γ-oryzanol (4000–11,000 ppm) for 4 weeks significantly reduced LDL-C levels compared with the control (Bumrungpert et al. 2019). The data further demonstrated that RBO containing γ-oryzanol could reduce LDL-C levels and elevate antioxidant capacity in hyperlipidemic subjects, suggesting that consumption of RBO may decrease CVD risk factors (Bumrungpert et al. 2019). Likewise, a pilot randomized controlled clinical trial involving children aged 8–13 years old with abnormal lipid profiles has demonstrated that consumption of 15 g of rice bran for 4 weeks are tolerable for children (Borresen et al. 2017). However, no association was observed between rice bran and TC, HDL-C, LDL-C, or TG (Borresen et al. 2017). This finding could be attributed to the short dietary intervention period (4 weeks) and a small sample size that used to observe the changes in serum lipids (Borresen et al. 2017). Notably, RBO has been demonstrated to increase the HDL-C and decrease TG and TC in hyperlipidemic and type 2 diabetes patients (Salar et al. 2016; Bumrungpert et al. 2019). Data from prospective, open-label dietary study showed that intakes of RBO and sesame oil (80:20) as cooking oil in mild-to-moderate hypertensive patients increased HDL-C (11%) and reduced LDL-C (27.5%) and blood pressure (mean arterial pressure (MAP) = 12.9%; diastolic blood pressure (DBP) = 13.5%; systolic blood pressure (SBP) = 12.8%) (Devarajan et al. 2016). Intriguingly, Vissers et al. (2000) found that a diet containing 2.1 g of sterols can also reduce LDL-C and TC levels in healthy individuals. Collectively, rice by-products protect against CVD by modulating lipid metabolism, regulating blood pressure, attenuating oxidative stress, and reducing inflammation may be modulated partly through the additive/synergistic effect of bioactive compounds.

5.4 Hypoglycemic Effect

Diabetes is a metabolic, chronic disease characterized by increased blood glucose, affecting nearly 422 million adults worldwide in 2014, particularly in middle- and low-income countries (World Health Organization 2019b). About 3.7 million deaths are attributed to diabetes and high blood glucose (World Health Organization 2019b). Diabetes mellitus is a progressive and complex disease that is accompanied by many complications including micro- and macrovascular damage, neuropathy, retinopathy, and nephropathy (Fowler 2011). Substantial evidence has revealed that hyperglycemia leads to the production of ROS and thereby increasing oxidative stress in several tissues (Zgheib et al. 2017). The redox imbalance is more likely induced in the absence of an appropriate compensatory response from the endogenous antioxidant network via stimulation of stress-sensitive intracellular pathways (Yerra et al. 2018). Type 2 diabetes is characterized by insulin resistance, reduced insulin secretion, and excessive hepatic glucose production (Czech 2017). Deterioration and impaired glucose tolerance are triggered when reduced insulin secretory responses or increased insulin resistance (Færch et al. 2016). The elevation of glucose levels produced oxidative stress due to the glucose autoxidation (Wolff and Dean 1987; Wolff et al. 1991), nonenzymatic glycation of proteins (Brownlee and Cerami 1981; Brownlee 2000), and increased production of mitochondrial ROS (Brownlee 2001).

The elevation of serum insulin levels along with the increase of blood glucose suggests that insulin action on glucose regulation was impaired after administration of a high carbohydrate-high fat diet (Boonloh et al. 2015). Feeding high carbohydrate-high fat diet rats with rice bran protein reduced homeostasis model assessment-insulin resistance (HOMA-IR) scores, serum insulin, and blood glucose levels (Boonloh et al. 2015) (Table 5.7). The favorable effect may be linked to the enhanced insulin action in peripheral tissue such as skeletal muscle, white adipose tissue, and liver. The improvement in lipid metabolism by rice bran protein may contribute to the alleviation of insulin resistance (Boonloh et al. 2015).

Adiponectin has been recognized as an insulin-sensitizing and anti-diabetic adipokine (Okamoto et al. 2006). It has been demonstrated that adiponectin reduces

Table 5.7 Animal studies conducted in rice by-products and their effects on diabetes

Animal	Rice by-products	Findings	References
Dairy cow	Neutral detergent fiber from rice straw	Reduced plasma glucose	Kanjanapruthipong and Thaboot (2006)
Obese rats	5% rice bran enzymatic extract	Ameliorates HOMA-IR levels and improved insulin sensitivity	Justo et al. (2013)
High carbohydrate-high fat diet rats	Rice bran protein	Reduced HOMA-IR scores, serum insulin, and blood glucose levels	Boonloh et al. (2015)
Insulin resistance rats	RBO	Improved insulin resistance	Abd El-Wahab et al. (2017)
Obese rats	GABA enriched rice bran	Exhibited a more efficient in reducing serum sphingolipids by decreasing transcriptional activity of *Ppp2r3b* and *Prkcg*	Si et al. (2018)
Fructose-fed rats	RBO	Improved insulin sensitivity and decreased HOMA-IR	Abd Elbast et al. (2018)
Weaned piglets	HEBR	No effect on blood glucose levels	Huang et al. (2019)

GABA γ-aminobutyric acid, *HEBR* hydrolysates produced by limited enzymatic broken rice, *HOMA-IR* homeostasis model assessment-insulin resistance, *RBO* rice bran oil

hepatic gluconeogenesis and promotes fatty acid oxidation and glucose uptake. Rice bran protein was found to elevate adiponectin levels in high carbohydrate-high fat-fed rats and thereby lowering blood glucose and enhancing insulin sensitivity (Boonloh et al. 2015). The upregulation of PPARγ enhances glucose metabolism and insulin sensitivity (Tsuchida et al. 2005; Viana Abranches et al. 2011). It has been suggested that PPARγ stimulation enhances the secretion of adipokines including adiponectin (Kim et al. 2008; Pereira et al. 2008). Since rice bran protein could promote PPARγ activation, it implied that rice bran protein may elevate the adiponectin secretion via activation of PPARγ. Intriguingly, high carbohydrate-high fat-fed rodent and obese humans showed an increase in proinflammatory adipokines levels, which are believed to trigger insulin resistance (Bastard et al. 2006; Raucci et al. 2013). Si et al. (2018) exploring the impact of GABA enriched rice bran on metabolic syndromes induced by a high-fat diet. The data showed that GABA enriched rice bran exhibited a more efficient in reducing serum sphingolipids, which is tightly linked to insulin resistance. A similar dietary supplementation was also demonstrated to reduce the transcriptional activity of *Prkcg* and *Ppp2r3b*, which is involved in the function of ceramides in blocking insulin activity (Si et al. 2018). Further, research evidence indicates that dietary fiber decreased blood glucose and suppressed inflammatory response (North et al. 2009). It has been demonstrated that a low-fiber diet can stimulate IL-6 expression and hyperglycemia (North et al. 2009). Intriguingly, insulin enhances the secretion of IL-6 levels in *in vitro* (Krogh-Madsen et al. 2004) and *in vivo* (Vicennati et al. 2002) studies. In this regard, diets that can decrease insulin secretion are linked to the low levels of IL-6 (North et al. 2009). In addition to the effects reported above, RBO has the potential to protect diabetes. An animal study has revealed that feeding rats with RBO improved insulin sensitivity and decreased HOMA-IR compared to the fructose-fed rats (Abd Elbast et al. 2018). Declining insulin levels along with the decrease levels of HOMA-IR and glucose suggested that RBO exerts hypoglycemic activity by improving insulin action. A similar dietary intake of RBO was also found to improve insulin resistance (Abd Elbast et al. 2016; Abd El-Wahab et al. 2017). Figure 5.4 shows the mechanisms of rice by-products involved in the modulation of diabetes.

Dietary administration of neutral detergent fiber from rice straw in dairy cows reduced plasma glucose (Kanjanapruthipong and Thaboot 2006). Propionic acid is the predominant precursor for hepatic glucose production (Danfaer et al. 1995). Intakes of high fiber diet produce low levels of propionic acid in the rumen (Schwartz and Gilchrist 1975). Reduced plasma glucose levels after dietary administration of roughage neutral detergent fiber from rice straw were due to the decreased gluconeogenesis from propionic acid produced in the rumen (Kanjanapruthipong and Thaboot 2006). However, administration of HEBR (30 g/kg for 21 days) in weaned piglets did not significantly different in blood glucose levels compared to control (Huang et al. 2019). Administration of 5% rice bran enzymatic extract ameliorates HOMA-IR levels in obese rats. In addition, the insulin sensitivity was also enhanced after feeding with 5% of rice bran enzymatic extract (Justo et al. 2013). Although

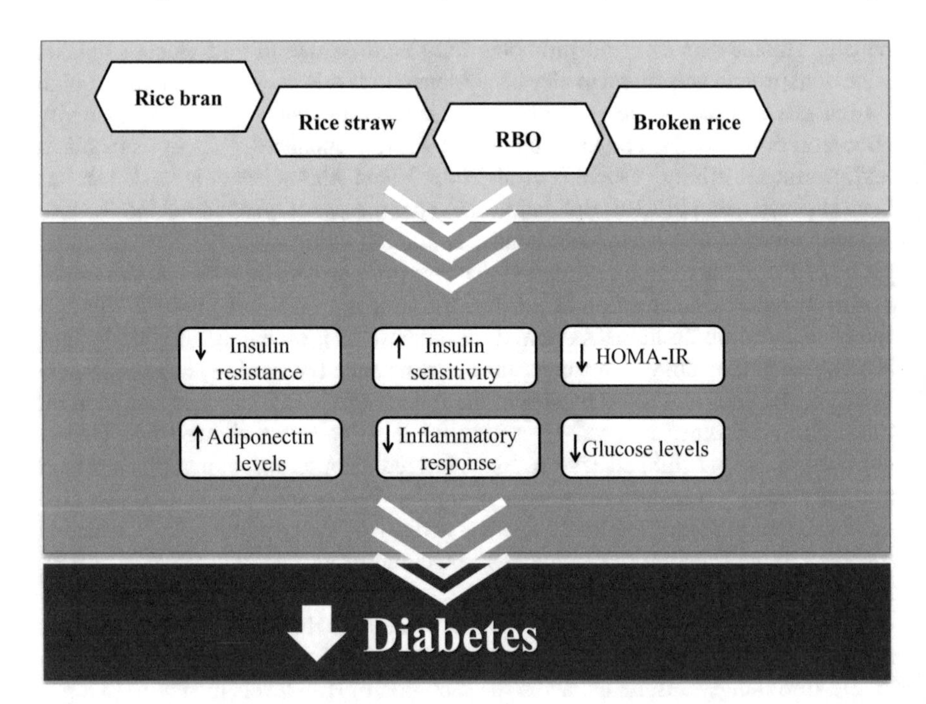

Fig. 5.4 The mechanisms of rice by-products involved in the modulation of diabetes. *HOMA-IR* homeostasis model assessment-insulin resistance, *RBO* rice bran oil

the extract did not alter fasting glucose, the data clearly indicate the beneficial effects of rice bran enzymatic extract on improving insulin sensitivity. In this regard, the protein derived from the enzymatic treatment in rice bran might be involved in the modulation of insulin metabolism and glucose levels, since this protein may decrease the formation of advanced glycosylation end products (AGEs) (Qureshi et al. 2002). AGEs are formed in high levels under a diabetic state (Qureshi et al. 2002). Furthermore, Qureshi et al. (2002) found that stabilized rice bran extract can effectively control insulin resistance and blood glucose levels in diabetic humans. The amelioration of complications in diabetes such as glycoxidation and glycation has been hypothesized due to the synergistic activity of several bioactive components, especially γ-oryzanol and tocopherols (Chou et al. 2009). Based on the evidence, rice by-products may reduce oxidative stress and alleviate insulin resistance. However, the potential implication of dietary intake of rice by-products in relation to diabetes is required to be elucidated further in long-term clinical studies.

5.5 Other Related Health Benefits

5.5.1 Neurodegenerative Diseases

Dementia is a syndrome that deteriorates behavior, thinking, and memory as well as the ability to conduct daily activities (World Health Organization 2019c). Dementia is a chronic or progressive condition, affecting nearly 50 million people worldwide. Alzheimer's disease is the most common condition of dementia, representing about 60–70% of cases (World Health Organization 2019c) (Fig. 5.5). Mitochondrial dysfunction has emerged as an initial event in the development of Alzheimer's disease and plays a vital role in brain aging (Stephanie et al. 2015).

The study suggests that food ingredients can enhance mitochondrial function (Fig. 5.5). The levels of γ-carboxyethyl hydroxychroman (CEHC) and α-tocotrienol were increased in brains of guinea pigs administered with stabilized rice bran extract; conversely, the concentration of γ-tocotrienol was not affected (Hagl et al. 2013) (Table 5.8). The mitochondrial and respiration coupling were significantly improved mitochondrial function in brains of stabilized rice bran extract fed animals, suggesting the potential of stabilized rice bran extract in the prevention of oxidative stress and mitochondrial dysfunction in neurodegenerative disease and brain aging (Hagl et al. 2013). Consistent with the findings demonstrated by Hagl et al. (2013), Stephanie et al. (2015) also found that stabilized rice bran extract promotes respiratory rates and ATP production and upregulates peroxisome proliferator-activated receptor gamma coactivator 1-alpha (PGC1α) protein expression in PC12APPsw cells, and thereby enhancing the impaired mitochondrial function. PC12APPsw cells are cell culture models of Alzheimer's disease, which exert very low levels of amyloid-β and impaired energy metabolism (Stephanie et al. 2015). Such finding highlights the unique components in rice bran extract; all these

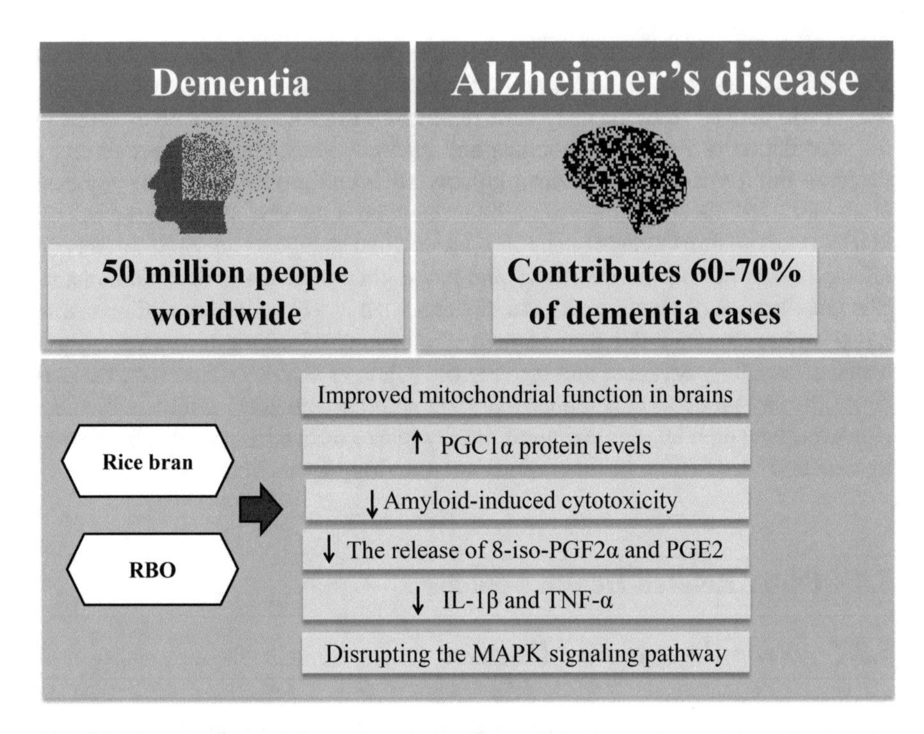

Fig. 5.5 The prevalence of dementia and the effects of rice by-products on neurodegenerative disease. *IL-1β* interleukin-1beta, *MAPK* mitogen-activated protein kinase, *PGC1α* peroxisome proliferator-activated receptor gamma coactivator 1-alpha, *PGE_2* prostaglandin E2, *RBO* rice bran oil, *TNF-α* tumor necrosis factor-alpha

Table 5.8 Effect of rice by-products on neurodegenerative diseases

Preclinical studies	Rice by-products	Findings	References
Neuroblastoma cells	Peptides derived from rice bran	Reduced amyloid-induced cytotoxicity	Kannan et al. (2012)
Guinea pigs	Stabilized rice bran extract	Enhanced the levels of CEHC and α-tocotrienol but no effect on the concentration of γ-tocotrienol	Hagl et al. (2013)
Naval Medical Research Institute mice	Rice bran extract	Improved aging-associated mitochondrial dysfunction	Hagl et al. (2015)
PC12APPsw cells	Stabilized rice bran extract	Promotes respiratory rates and ATP production and upregulates PGC1α protein levels	Stephanie et al. (2015)
Primary rat microglia	Rice bran extract	Inhibited the release of 8-iso-PGF2α and PGE2 levels in LPS-activated primary microglia	Bhatia et al. (2016)
Wistar rats	RBO	Ameliorated the decline in spatial memory	Kumar et al. (2019)

ATP adenosine triphosphate, *CEHC* γ-carboxyethyl hydroxychroman, *LPS* lipopolysaccharides, *PGC1α* peroxisome proliferator-activated receptor gamma coactivator 1-alpha, *PGE_2* prostaglandin E2, *RBO* rice bran oil substances have shown beneficial effects on mitochondrial function in the early stage of Alzheimer's disease, as demonstrated in a cell culture model (Stephanie et al. 2015). In another study, Kannan et al. (2012) showed that peptides derived from rice bran can reduce amyloid-induced cytotoxicity in neuroblastoma cells.

Hyperactivation of microglia plays an important role in modulating neuroinflammatory activities and is considered as a hallmark of brain inflammation (Krabbe et al. 2017). Indeed, neuroinflammatory changes in microglia are one of the key risk factor for neurodegenerative disease. Oral administration of rice bran extract for 3 weeks improved aging-associated mitochondrial dysfunction in 18-month-old Naval Medical Research Institute (NMRI) mice (Hagl et al. 2015). An animal study showed that rice bran extract inhibits the release of 8-iso-$PGF_{2\alpha}$ and prostaglandin (PG) E_2 levels in LPS-activated primary microglia. 8-iso-$PGF_{2\alpha}$ is a biomarker for free radical formation of lipid hydroperoxides. Notably, treatment activated microglia with rice bran extract downregulated proinflammatory M1 expression (IL-1β and TNF-α) and upregulated transcriptional activity of microglial M2 marker IL-10. The data further demonstrated that rice bran extract suppressed microglial activation by disrupting the MAPK signaling pathway, implied that rice bran extract may prevent microglial dysfunction associated with neuroinflammatory diseases such as Alzheimer's disease (Bhatia et al. 2016). Indeed, increased proliferation of microglia could be partly due to the synergistic effect of bioactive constituents in the rice bran extract, such as tocopherols or tocotrienols. In support of this, Flanary and Streit (2006) and Ren et al. (2010) found that tocotrienols and α-tocopherol trigger the proliferation in cultured rat microglia. High intakes of n-3 polyunsaturated fatty acids (PUFA) are negatively linked to Alzheimer's disease risk. A study by Kumar et al. (2019) found that RBO ameliorates the decline in spatial memory induced by aluminium chloride ($AlCl_3$) in Wistar rats, suggesting that unsaturated fats in RBO may protect against dementia. Due to the bioactive compounds in rice by-products, there has been a tremendous interest in the study of antioxidant activity and its components on neurodegenerative disease. Together, the modulation of microglia activation played by rice by-products is nonetheless worth study in long-term animal studies.

5.5.2 *Osteoporosis*

Osteoclast has been implicated in the development of metabolic disorders such as osteoporosis (Zha et al. 2018). The implication is mainly due to the imbalance induced by decreased osteoblastic bone-forming activity and increased osteoclastic bone resorptive activity (Boyle et al. 2003). Bone homeostasis is maintained when there is a close balance between osteoclast-related bone resorption and osteoblast-related bone formation (Chen et al. 2018). In this regard, several strategies are adopted to decrease the incidence of osteoporosis, including induction of bone formation using anabolic agents such as parathyroid hormone (PTH) and reduction of bone resorption using anti-resorptive agent including bisphosphonates. Anti-resorptive

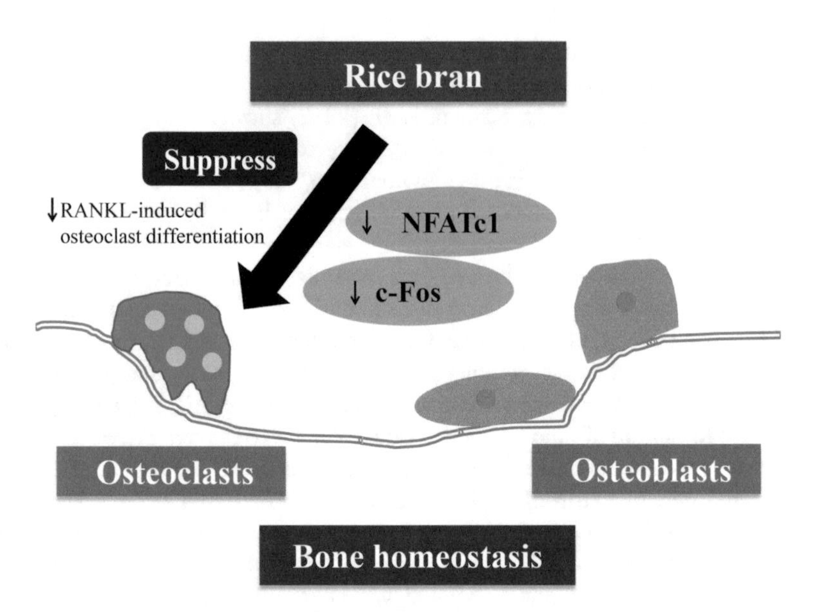

Fig. 5.6 The interaction of rice bran in relation to osteoclasts. *NFATc1* nuclear factor of activated T-cells, cytoplasmic 1, *RANKL* receptor activator of nuclear factor κB ligand

agents such as bisphosphonates are one of the most common therapeutic agents for osteoporosis; however, this strategy appears to cause an adverse outcome, for instance, atypical femoral fracture (Meier et al. 2012) and bisphosphonate-associated osteonecrosis of the jaw (Dannemann et al. 2007). Despite anabolic PTH is used for activating bone formation, the used is limited by its long-term safety and costs. Therefore, the anti-osteoporotic agent from natural product-derived extract has drawn a great deal of interest. Rice bran exerts multiple biological activities in the treatment and prevention of metabolic disorders (Henderson et al. 2012). Among all the bioactive compounds in rice bran, several components such as tocotrienols and tocopherols, hydroxycinnamic acid, and ferulic acid have been demonstrated to prevent bone loss or suppress osteoclastogenesis (Ha et al. 2011; Shuid et al. 2019) (Fig. 5.6). Butanol extract of rice bran has the potential to suppress receptor activator of nuclear factor κB ligand (RANKL)-induced osteoclast differentiation (Table 5.9). The ability of butanol extract of rice bran to alleviate the RANKL-induced stimulations of MAPK pathway could inhibit the inductions of c-Fos and nuclear factor of activated T-cells, cytoplasmic 1 (NFATc1) levels that downregulate the expressions of NFATc1-controlled osteoclast-specific genes, for instance, dendritic cell-specific transmembrane protein (DC-STAMP) and cathepsin K (Moon et al. 2013), which are essential for bone resorption and osteoclast fusion (Ishikawa et al. 2001; Asagiri et al. 2005), suggesting that butanol extract of rice bran may have the potential to suppress osteoclastogenesis, resorptive activity, and osteoclast fusion (Fig. 5.6). A study by Heli Roy and Shanna Lundy (2005) further demonstrated the unique complex of bioactive compounds of rice bran. Strikingly, diet containing rice bran oryzanol is negatively linked to bone loss in women who suffered from postmenopausal osteoporosis (Heli

Table 5.9 Effect of rice by-products on osteoporosis and arthritis

Studies	Rice by-products	Findings	References
Osteoporosis			
Older women with osteoporosis	Rice bran	Improved bone loss	Colona (2002)
Postmenopausal osteoporosis women	Rice bran oryzanol	Negatively linked to bone loss	Heli Roy and Shanna Lundy (2005)
Male ICR mice	Butanol extract of rice bran	Suppressed RANKL-induced osteoclast differentiation	Moon et al. (2013)
Arthritis			
–	Stabilized rice bran	Inhibits 5-LOX, COX-2, and COX-1 enzymes	Roschek et al. (2009)
Adjuvant-induced arthritis in rats	RBO	Ameliorates cytokines and inflammatory eicosanoids, decreases oxidative stress, and alleviates the severity of paw inflammation	Yadav et al. (2016)

COX-1 cyclooxygenase-1, *COX-2* cyclooxygenase-2, *ICR* imprinting control region, *RANKL* receptor activator of nuclear factor κB ligand, *RBO* rice bran oil, *5-LOX* 5-lipooxygenase

Roy and Shanna Lundy 2005). Rice bran can also improve bone loss in older women who suffered from osteoporosis (Colona 2002). Besides its effects on osteoporosis, the previous study has also been reported that rice bran can be used as ergogenic supplements by athletes and bodybuilders (Fry et al. 1997), demonstrating the tremendous functional potential of rice bran. Overall, rice bran could be considered as a potential treatment for osteoclast-associated disorders. Further study is warranted to identify and evaluate the anti-resorptive activity in mature osteoclasts.

5.5.3 Arthritis

Musculoskeletal condition is a major contributor to morbidity and disability. It can be classified as physical disability, spinal disorders, and joint diseases (World Health Organization 2019d). Arthritis has been identified as an inflammatory disorder (Nerurkar et al. 2019). The interaction between oxidative stress and inflammation is closely linked to the PGs biosynthetic pathway that generates reactive species (Kawahara et al. 2015). Cyclooxygenase (COX) is a key enzyme in the synthesis of PG, in which the arachidonic acid produces prostaglandin H_2 (PGH_2) (Shehzad et al. 2015). COX-2 and PGE_2 levels are markedly increased in rheumatoid synovium compared to less inflamed osteoarthritic synovium (Kitano et al. 2006; Benito et al. 2005). In this regard, COX-2 selective inhibitors are potential targets for inflammatory response and arthritis to modulate the downstream proinflammatory PGs. Besides COX-2 expression, there is considerable evidence revealed that 5-lipoxygenase (5-LOX) plays a vital role in proinflammatory pathways (Martel-Pelletier et al. 2003;

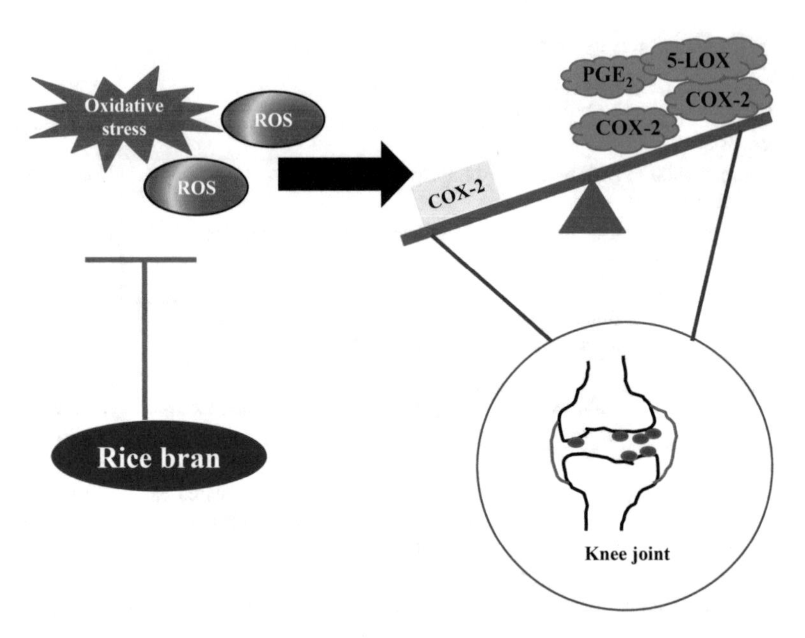

Fig. 5.7 The effect of dietary intake of rice bran in relation to oxidative stress in arthritis. *COX-2* cyclooxygenase-2, *PGE_2* prostaglandin E2, *ROS* reactive oxygen species, *5-LOX* 5-lipoxygenase

Sharma and Mohammed 2006; Whitehouse and Rainsford 2006). 5-LOX produces more than four leukotrienes, namely E_4, D_4, C_4, and B_4, and cytokines, which are subsequently led to bone resorption and joint inflammation (Martel-Pelletier et al. 2003). Roschek et al. (2009) found that stabilized rice bran inhibits 5-LOX, COX-2, and COX-1 enzymes (Table 5.9). The fatty acids identified in the stabilized rice bran extract are likely nonselective inhibitors of LOX and COX enzymes. The effect is similar to omega-3 PUFA and fish oils, which inhibit the incorporation of the active site substrates linoleic acid and arachidonic acid into the corresponding enzymes and thereby suppressing the production of leukotrienes and inflammatory PGs. Intriguingly, the alkaloids including acetyllaburnine and epiloliolide, and phenolics such as 12-shogaol, are likely selectively COX-2 inhibitors. The molecule regions, especially phenolic/aromatic, are similar to COX-2 inhibitors and thus contribute to the COX-2 selectivity (Roschek et al. 2009). RBO reduces the severity of adjuvant-induced arthritis in rats through ameliorating cytokines and inflammatory eicosanoids, decreasing oxidative stress, and alleviating the severity of paw inflammation (Yadav et al. 2016). Overall, the bioactive constituents in rice bran or RBO possess an anti-arthritic activity that is responsible for decreasing inflammation. Figure 5.7 summarizes the effect of dietary intake of rice bran on arthritis.

References

Abd Elbast SA, Rashed LA, Mohamed MA et al (2016) Amelioration of insulin resistance in rats treated with rice bran oil. Egyp J Hosp Med 65:547–552

Abd Elbast SA, Mohamed MA, Ahmed MA et al (2018) Rice bran oil ameliorates hepatic insulin resistance in fructose fed-rats. Egypt J Hosp Med 71:2885–2891

Abd El-Wahab HMF, Mohamed MA, El Sayed HH et al (2017) Modulatory effects of rice bran and its oil on lipid metabolism in insulin resistance rats. J Food Biochem 41:12318

Abdollahi M, Afshar-Imani B (2003) A review on obesity and weight loss measures. Middle East Pharm 11:6–10

Abete I, Goyenechea E, Zulet MA et al (2011) Obesity and metabolic syndrome: potential benefit from specific nutritional components. Nutr Metab Cardiovasc Dis 21:B1–B15

Abraham A, Kattoor AJ, Saldeen T et al (2019) Vitamin E and its anticancer effects. Crit Rev Food Sci Nutr 59:2831–2838

Ajuluchukwu JN, Okubadejo NU, Mabayoje M et al (2007) Comparative study of the effect of tocotrienols and -tocopherol on fasting serum lipid profiles in patients with mild hypercholesterolaemia: a preliminary report. Niger Postgrad Med J 14:30–33

Anand P, Kunnumakkara AB, Sundaram C et al (2008) Cancer is a preventable disease that requires major lifestyle changes. Pharm Res 25:2097–2116

Asagiri M, Sato K, Usami T et al (2005) Autoamplification of NFATc1 expression determines its essential role in bone homeostasis. J Exp Med 202:1261–1269

Baliarsingh S, Beg ZH, Ahmad J (2005) The therapeutic impacts of tocotrienols in type 2 diabetic patients with hyperlipidemia. Atherosclerosis 182:367–374

Bang MH, Van Riep T, Thinh NT et al (2010) Arabinoxylan rice bran (MGN-3) enhances the effects of interventional therapies for the treatment of hepatocellular carcinoma: a three-year randomized clinical trial. Anticancer Res 30:5145–5151

Barish GD, Narkar VA, Evans RM (2006) PPARδ: a dagger in the heart of the metabolic syndrome. J Clin Investig 116:590–597

Bastard JP, Maachi M, Lagathu C et al (2006) Recent advances in the relationship between obesity, inflammation, and insulin resistance. Eur Cytokine Netw 17:4–12

Benito MJ, Veale DJ, FitzGerald O et al (2005) Synovial tissue inflammation in early and late osteoarthritis. Ann Rheum Dis 64:1263–1267

Bhatia HS, Baron J, Hagl S et al (2016) Rice bran derivatives alleviate microglia activation: possible involvement of MAPK pathway. J Neuroinflammation 13:148

Billadeau DD (2007) Primers on molecular pathways: the glycogen synthase kinase-3β. Pancreatology 7:398–402

Boonloh K, Kukongviriyapan V, Kongyingyoes B et al (2015) Rice bran protein hydrolysates improve insulin resistance and decrease pro-inflammatory cytokine gene expression in rats fed a high carbohydrate-high fat diet. Nutrients 7:6313–6329

Borresen EC, Jenkins-Puccetti N, Schmitz K et al (2017) A pilot randomized controlled clinical trial to assess tolerance and efficacy of navy bean and rice bran supplementation for lowering cholesterol in children. Glob Pediatr Health 4:1–10

Bouchard C (2008) Gene-environment interactions in the etiology of obesity: defining the fundamentals. Obesity (Silver Spring) 16:S5–S10

Boyle WJ, Simonet WS, Lacey DL (2003) Osteoclast differentiation and activation. Nature 423:337–342

Brauner R, Johannes C, Ploessl F et al (2012) Phytosterols reduce cholesterol absorption by inhibition of 27-hydroxycholesterol generation, liver X receptor a activation, and expression of the basolateral sterol exporter ATP-binding cassette A1 in Caco-2 enterocytes. J Nutr 142:981–989

Brown DG, Borresen EC, Brown RJ et al (2017) Heat-stabilised rice bran consumption by colorectal cancer survivors modulates stool metabolite profiles and metabolic networks: a randomised controlled trial. Br J Nutr 117:1244–1256

Brownlee M (2000) Negative consequences of glycation. Metabolism 49:9–13

Brownlee M (2001) Biochemistry and molecular cell biology of diabetic complications. Nature 414:813–820
Brownlee M, Cerami A (1981) The biochemistry of the complications of diabetes mellitus. Annu Rev Biochem 50:385–432
Bumrungpert A, Chongsuwat R, Phosat C et al (2019) Rice bran oil containing gamma-oryzanol improves lipid profiles and antioxidant status in hyperlipidemic subjects: a randomized double-blind controlled trial. J Altern Complement Med 25:353–358
Chang H, Lei L, Zhou Y et al (2018) Dietary flavonoids and the risk of colorectal cancer: an updated meta-analysis of epidemiological studies. Nutrients 10:950
Chaput JP, Klingenberg L, Astrup A et al (2011) Modern sedentary activities promote overconsumption of food in our current obesogenic environment. Obes Rev 12:e12–e20
Chen M-H, Choi SH, Kozukue N et al (2012) Growth-inhibitory effects of pigmented rice bran extracts and three red bran fractions against human cancer cells: relationships with composition and antioxidative activities. J Agric Food Chem 60:9151–9161
Chen X, Wang Z, Duan N et al (2018) Osteoblast-osteoclast interactions. Connect Tissue Res 59:99–107
Cholujova D, Jakubikova J, Czako B et al (2013) MGN-3 arabinoxylan rice bran modulates innate immunity in multiple myeloma patients. Cancer Immunol Immunother 62:437–445
Chou TW, Ma CY, Cheng HH et al (2009) A rice bran oil diet improves lipid abnormalities and suppress hyperinsulinemic responses in rats with streptozotocin/nicotinamide-induced type 2 diabetes. J Clin Biochem Nutr 45:29–36
Cicero AF, Gaddi A (2001) Rice bran oil and gamma-oryzanol in the treatment of hyperlipoproteinaemias and other conditions. Phytother Res 15:277–289
Colona HC (2002) The effects of oryzanol on bone mineral density in ovariectomized, retired breeder rats. Thesis, Louisiana State University
Czech MP (2017) Insulin action and resistance in obesity and type 2 diabetes. Nat Med 23:804–814
Danfaer A, Tetens V, Agergaard N (1995) Review and an experimental study on the physiological and quantitative aspects of gluconeogenesis in lactating ruminants. Comp Biochem Physiol 111B:201–210
Dannemann C, Gratz KW, Riener MO et al (2007) Jaw osteonecrosis related to bisphosphonate therapy: a severe secondary disorder. Bone 40:828–834
Davi G, Santilli F, Patrono C (2010) Nutraceuticals in diabetes and metabolic syndrome. Cardiovasc Ther 28:216–226
de Heredia FP, Gómez-Martinez S, Marcos A (2012) Obesity, inflammation and the immune system. Proc Nutr Soc 71:332–338
Devarajan S, Singh R, Chatterjee B et al (2016) A blend of sesame oil and rice bran oil lowers blood pressure and improves the lipid profile in mild-to-moderate hypertensive patients. J Clin Lipidol 10:339–349
Edrisi F, Salehi M, Ahmadi A et al (2018) Effects of supplementation with rice husk powder and rice bran on inflammatory factors in overweight and obese adults following an energy-restricted diet: a randomized controlled trial. Eur J Nutr 67:833–843
Egeberg R, Olsen A, Loft S et al (2010) Intake of wholegrain products and risk of colorectal cancers in the Diet, Cancer and Health cohort study. Br J Cancer 103:730–734
El-Din NKB, Ali DA, El-Dein MA et al (2016) Enhancing the apoptotic effect of a low dose of paclitaxel on tumor cells in mice by arabinoxylan rice bran (MGN-3/Biobran). Nutr Cancer 68:1010–1020
El-Din NKB, Areida SK, Ahmed KO et al (2019) Arabinoxylan rice bran (MGN-3/Biobran) enhances radiotherapy in animals bearing Ehrlich ascites carcinoma. J Radiat Res 60:747–758
Erlinawati ND, Oetoro S, Gunarti DR (2017) Effect of rice bran oil on the lipid profile of mild-moderate hypercholesterolemic male aged 19–55 year. World Nutr J 1:52–57
Esa NM, Ling TB, Peng LS (2013) By-products of rice processing: an overview of health benefits and applications. J Rice Res 1:107

Evans CEL (2019) Dietary fibre and cardiovascular health: a review of current evidence and policy. Proceedings of the Nutrition Society. Cambridge University Press

Expert Panel Members, Jensen MD, Ryan DH et al (2013) Executive summary: guidelines (2013) for the management of overweight and obesity in adults: a report of the American College of Cardiology/American Heart Association task force on practice guidelines and the obesity society published by the Obesity Society and American College of Cardiology/American Heart Association task force on practice guidelines. Based on a systematic review from the Obesity Expert Panel. Obesity (Silver Spring) 22:S5–S39

Faam BZM, Daneshpour M, Azizi F et al (2014) Association between abdominal obesity and hs-CRP, IL-6 and HCY in tehranian adults: TLGS. J Diabetes Metab Disord 13:163–171

Færch K, Vistisen D, Pacini G et al (2016) Insulin resistance is accompanied by increased fasting glucagon and delayed glucagon suppression in individuals with normal and impaired glucose regulation. Diabetes 65:3473–3481

Ferdinand O, Sen B, Rahurkar S et al (2012) The relationship between built environments and physical activity: a systematic review. Am J Public Health 102:e7–e13

Fiaschi T, Chiarugi P (2012) Oxidative stress, tumor microenvironment, and metabolic reprogramming: a diabolic liaison. Int J Cell Biol 2012:762825. 8p

Flanary BE, Streit WJ (2006) Alpha-tocopherol (vitamin E) induces rapid, nonsustained proliferation in cultured rat microglia. Glia 53:669–674

Forster GM, Raina K, Kumar A et al (2013) Rice varietal differences in bioactive bran components for inhibition of colorectal cancer cell growth. Food Chem 141:1545–1552

Fowler MJ (2011) Microvascular and macrovascular complications of diabetes. Clin Diab 29:116–122

Fry AC, Bonner E, Lewis DL et al (1997) The effects of gamma-oryzanol supplementation during resistance exercise training. Int J Sport Nutr 7:318–329

Fung TT, Hu FB, Wu K et al (2010) The Mediterranean and Dietary Approaches to Stop Hypertension (DASH) diets and colorectal cancer. Am J Clin Nutr 92:1429–1435

Galisteo M, Duarte J, Zarzuelo A (2008) Effects of dietary fibers on disturbances clustered in the metabolic syndrome. J Nutr Biochem 19:71–84

Gallaher DD, Gallaher CM, Mahrt GJ et al (2002) A glucomannan and chitosan fiber supplement decreases plasma cholesterol and increases cholesterol excretion in overweight normocholesterolemic humans. J Am Coll Nutr 21:428–433

Gholamian-Dehkordi N, Luther T, Asadi-Samani M et al (2017) An overview on natural antioxidants for oxidative stress reduction in cancers; a systematic review. Immunopathol Persa 3:e12

Ghoneum M, Brown J (1999) NK immunorestoration of cancer patients by BioBran/MGN-3, a modified arabinoxylan rice bran (study of 32 patients followed for up to 4 years). Anti-Aging Med Ther 3:217–226

Grooms KN, Ommerborn MJ, Pham DQ et al (2013) Dietary fiber intake and cardiometabolic risks among US adults, NHANES 1999–2010. Am J Med 126:1059–1067

Grummer RR, Carroll DJ (1988) A review of lipoprotein cholesterol metabolism: importance to ovarian function. J Anim Sci 66:3160–3173

Gunasekaran S, Venkatachalam K, Namasivayam N (2015) p-Methoxycinnamic acid, an active phenylpropanoid induces mitochondrial mediated apoptosis in HCT-116 human colon adenocarcinoma cell line. Environ Toxicol Pharmacol 40:966–974

Gupta SC, Hevia D, Patchva S et al (2012) Upsides and downsides of reactive oxygen species for cancer: the roles of reactive oxygen species in tumorigenesis, prevention, and therapy. Antioxid Redox Signal 16:1295–1322

Gysin R, Azzi A, Visarius T (2002) Gamma-tocopherol inhibits human cancer cell cycle progression and cell proliferation by down-regulation of cyclins. FASEB J 16:1952–1954

Ha H, Lee JH, Kim HN et al (2011) α-Tocotrienol inhibits osteoclastic bone resorption by suppressing RANKL expression and signaling and bone resorbing activity. Biochem Biophys Res Commun 406:546–551

Hagl S, Kocher A, Schiborr C et al (2013) Rice bran extract protects from mitochondrial dysfunction in guinea pig brains. Pharmacol Res 76:17–27
Hagl S, Berressem D, Grewal R et al (2015) Rice bran extract improves mitochondrial dysfunction in brains of aged NMRI mice. Nutr Neurosci 19:1–10
Hanahan D, Weinberg RA (2011) Hallmarks of cancer: the next generation. Cell 144:646–674
He T-C, Sparks AB, Rago C et al (1998) Identifcation of c-MYC as a target of the APC pathway. Science 281:1509–1512
Heli Roy RD, Shanna Lundy BS (2005) Rice bran. Pennington Nutrition Series 8. https://www.pbrc.edu/training-and-education/pdf/pns/PNS_Ricebran.pdf. Accessed 29 Dec 2019
Henderson AJ, Ollila CA, Kumar A et al (2012) Chemopreventive properties of dietary rice bran: current status and future prospects. Adv Nutr 3:643–653
Hink U, Li HG, Mollnau H et al (2001) Mechanisms underlying endothelial dysfunction in diabetes mellitus. Circ Res 88:E14–E22
Hotamisligil GS (2006) Inflammation and metabolic disorders. Nature 444:860–867
Huang Z, Peng H, Sun Y et al (2019) Beneficial effects of novel hydrolysates produced by limited enzymatic broken rice on the gut microbiota and intestinal morphology in weaned piglets. J Funct Foods 62:103560
Hullar MA, Burnett-Hartman AN, Lampe JW (2014) Gut microbes, diet, and cancer. Cancer Treat Res 159:377–399
Ijiri D, Nojima T, Kawaguchi M et al (2015) Effects of feeding outer bran fraction of rice on lipid accumulation and fecal excretion in rats. Biosci Biotechnol Biochem 79:1337–1341
Ishikawa T, Kamiyama M, Tani-Ishii N et al (2001) Inhibition of osteoclast differentiation and bone resorption by cathepsin K antisense oligonucleotides. Mol Carcinog 32:84–91
Ito Y, Nakashima Y, Matsuoka S (2015) Rice bran extract containing acylated steryl glucoside fraction decreases elevated blood LDL cholesterol level in obese Japanese men. J Med Investig 62:80–84
Itoh Y, Mizuno M, Ikeda M et al (2015) A randomized, double-blind pilot trial of hydrolyzed rice bran versus placebo for radioprotective effect on acute gastroenteritis secondary to chemoradiotherapy in patients with cervical cancer. Evid Based Complement Alternat Med 2015:974390. 6p
Jacobs DR Jr, Tapsell LC (2007) Food, not nutrients, is the fundamental unit in nutrition. Nutr Rev 65:439–450
Johansson-Persson A, Ulmius M, Cloetens L et al (2014) A high intake of dietary fiber influences C-reactive protein and fibrinogen, but not glucose and lipid metabolism, in mildly hypercholesterolemic subjects. Eur J Nutr 53:39–48
Justo ML, Rodriguez-Rodriguez R, Claro CM et al (2013) Water-soluble rice bran enzymatic extract attenuates dyslipidemia, hypertension and insulin resistance in obese Zucker rats. Eur J Nutr 52:789–797
Justo ML, Claro C, Zeyda M et al (2016) Rice bran prevents high-fat diet-induced inflammation and macrophage content in adipose tissue. Eur J Nutr 55:2011–2019
Kanjanapruthipong J, Thaboot B (2006) Effects of neutral detergent fiber from rice straw on blood metabolites and productivity of dairy cows in the tropics. Asian Australas J Anim Sci 19:356–362
Kannan A, Hettiarachchy NS, Mahedevan M (2012) Peptides derived from rice bran protect cells from obesity and Alzheimer's disease. Int J Biomed Res 3:131–135
Kawabata K, Tanaka T, Murakami T et al (1999) Dietary prevention of azoxymethane-induced colon carcinogenesis with rice-germ in F344 rats. Carcinogenesis 20:2109–2115
Kawahara K, Hohjoh H, Inazumi T et al (2015) Prostaglandin E 2-induced inflammation: relevance of prostaglandin E receptors. Biochim Biophys Acta 1851:414–421
Keys A, Anderson JT, Grande F (1960) Diet-type (fats constant) and blood lipids in man. J Nutr 70:257–266
Kim S-J, Park H-R, Park E et al (2007) Cytotoxic and antitumor activity of momilactone B from rice hulls. J Agric Food Chem 55:1702–1706

Kim HJ, Kim SK, Shim WS et al (2008) Rosiglitazone improves insulin sensitivity with increased serum leptin levels in patients with type 2 diabetes mellitus. Diabetes Res Clin Pract 81:42–49
Kim J-Y, Shin M, Heo Y-R (2014) Effects of stabilized rice bran on obesity and antioxidative enzyme activity in high fat diet-induced obese C57BL/6 mice. J Korean Soc Food Sci Nutr 43:1148–1157
King DE (2005) Dietary fiber, inflammation, and cardiovascular disease. Mol Nutr Food Res 49:594–600
Kitano M, Hla T, Sekiguchi M et al (2006) Sphingosine 1-phosphate/sphingosine 1-phosphate receptor 1 signaling in rheumatoid synovium: regulation of synovial proliferation and inflammatory gene expression. Arthritis Rheum 54:742–753
Knight ZA, Hannan KS, Greenberg ML et al (2010) Hyperleptinemia is required for the development of leptin resistance. PLoS One 5:e11376
Kojima S, Nakayama K, Ishida H (2004) Low dose γ -rays activate immune functions via induction of glutathione and delay tumor growth. J Radiat Res (Tokyo) 45:33–39
Krabbe G, Minami S, Etchegaray JI et al (2017) Microglial NFκB-TNFα hyperactivation induces obsessive–compulsive behavior in mouse models of progranulin-deficient frontotemporal dementia. Proc Natl Acad Sci USA 114:5029–5034
Kritika S, Virginia P (2018) Hypolipidemic actions of oat, legumes, barley and rice bran. Int J Adv Agric Sci Technol 5:1–6
Krogh-Madsen R, Plomgaard P, Keller P et al (2004) Insulin stimulates interleukin-6 and tumor necrosis factor-α gene expression in human subcutaneous adipose tissue. Am J Physiol Endocrinol Metab 286:E234–E238
Kumar A, Mallik SB, Rijal S et al (2019) Dietary oils ameliorate aluminum chloride-induced memory deficit in wistar rats. Pharmacogn Mag 15:36–42
Kuno T, Takahashi S, Tomita H et al (2015) Preventive effects of fermented brown rice and rice bran against N-nitrosobis (2-oxopropyl) amine-induced pancreatic tumorigenesis in male hamsters. Oncol Lett 10:3377–3384
Lane ML, Vesely DL (2013) Reduction of leptin levels by four cardiac hormones: implications for hypertension in obesity. Exp Ther Med 6:611–615
Leardkamolkarn V, Tongthep W, Suttiarporn P et al (2011) Chemopreventive properties of the bran extracted from a newly-developed Tai rice: the Riceberry. Food Chem 125:978–985
Lee SC, Chung I-M, Jin YJ et al (2008) Momilactone B, an allelochemical of rice hulls, induces apoptosis on human lymphoma cells (Jurkat) in a micromolar concentration. Nutr Cancer 60:542–551
Libby P, Ridker PM, Hansson GK et al (2009) Inflammation in atherosclerosis from pathophysiology to practice. J Am Coll Cardiol 54:2129–2138
Maingrette F, Renier G (2003) Leptin increases lipoprotein lipase secretion by macrophages: involvement of oxidative stress and protein kinase C. Diabetes 52:2121–2128
Malvicini M, Gutierrez-Moraga A, Rodriguez MM et al (2018) A tricin derivative from *Deschampsia antarctica* Desv. inhibits colorectal carcinoma growth and liver metastasis through the induction of a specific immune response. Mol Cancer Ther 17:966–976
Martel-Pelletier J, Lajeunesse D, Reboul P et al (2003) Therapeutic role of dual inhibitors of 5-LOX and COX, selective and non-selective non-steroidal anti-inflammatory drugs. Ann Rheum Dis 62:501–509
Matheson HB, Colón IS, Story JA (1995) Cholesterol 7 alpha-hydroxylase activity is increased by dietary modification with psyllium hydrocolloid, pectin, cholesterol and cholestyramine in rats. J Nutr 125:454–458
Meier RPH, Perneger TV, Stern R et al (2012) Increasing occurrence of atypical femoral fractures associated with bisphosphonate use. Arch Intern Med 172:930–936
Meselhy KM, Shams MM, Sherif NH et al (2018) Phytochemical study, potential cytotoxic and antioxidant activities of selected food byproducts (Pomegranate peel, rice bran, rice straw & Mulberry bark). Nat Prod Res 6:1–4

Meselhy KM, Shams MM, Sherif NH et al (2019) Phenolic profile and *in vivo* cytotoxic activity of rice straw extract. Pharmacogn J 11:849–857

Moon J, Moon S-H, Choi S-W et al (2013) Anti-osteoclastogenic activity of butanol fraction of rice bran extract via downregulation of MAP kinase activity and c-Fos/NFATc1 expression. Int J Med Med Sci 5:348–355

Mori H, Kawabata K, Yoshimi N et al (1999) Chemopreventive effects of ferulic acid on oral and rice germ on large bowel carcinogenesis. Anticancer Res 19:3775–3778

Nagasaka R, Yamsaki T, Uchida A et al (2011) γ-oryzanol recovers mouse hypoadiponectinemia induced by animal fat ingestion. Phytomedicine 18:669–671

Nandi I, Sengupta A, Ghosh M (2019) Effects of dietary fibres extracted from defatted sesame husk, rice bran and flaxseed on hypercholesteromic rats. Bioact Carbohydr Diet Fibre 17:100176

Nerurkar L, Siebert S, Mclnnes IB et al (2019) Rheumatoid arthritis and depression: an inflammatory perspective. Lancet Psychiatry 6:164–173

Newman RK, Betschart AA, Newman CW et al (1992) Effect of full-fat or defatted rice bran on serum cholesterol. Plant Foods Hum Nutr 42:37–43

Ng LT, Ko HJ (2012) Comparative effects of tocotrienol-rich fraction, α-tocopherol and α-tocopheryl acetate on inflammatory mediators and nuclear factor κB expression in mouse peritoneal macrophages. Food Chem 134:920–925

Norris L, Malkar A, Horner-Glister E et al (2015) Search for novel circulating cancer chemopreventive biomarkers of dietary rice bran intervention in Apc(Min) mice model of colorectal carcinogenesis, using proteomic and metabolic profiling strategies. Mol Nutr Food Res 59:1827–1836

North C, Venter C, Jerling J (2009) The effects of dietary fibre on C-reactive protein, an inflammation marker predicting cardiovascular disease. Eur J Clin Nutr 63:921–933

Oakes ND, Kjellstedt A, Thalén P et al (2013) Roles of fatty acid oversupply and impaired oxidation in lipid accumulation in tissues of obese rats. J Lipids 2013:420754. 12p

Ohara Y, Peterson TE, Harrison DG (1993) Hypercholesterolemia increases endothelial superoxide anion production. J Clin Invest 91:2546–2551

Okamoto Y, Kihara S, Funahashi T et al (2006) Adiponectin: a key adipocytokine in metabolic syndrome. Clin Sci (Lond) 110:267–278

Okonogi S, Kaewpinta A, Junmahasathien T et al (2018) Effect of rice variety and modification on antioxidant and anti-inflammatory activities. Drug Discov Ther 12:206–213

Ougolkov AV, Billadeau DD (2006) Targeting GSK-3: a promising approach for cancer therapy? Future Oncol 2:91–100

Parrado J, Miramontes E, Jover M et al (2003) Prevention of brain protein and lipid oxidation elicited by a water-soluble oryzanol enzymatic extract derived from rice bran. Eur J Nutr 42:307–314

Parrado J, Miramontes E, Jover M et al (2006) Preparation of a rice bran enzymatic extract with potential use as functional food. Food Chem 98:742–748

Payne CM, Bernstein C, Dvorak K et al (2008) Hydrophobic bile acids, genomic instability, Darwinian selection, and colon carcinogenesis. Clin Exp Gastroenterol 1:19–47

Pearce BC, Parker RA, Deason ME et al (1992) Hypocholesterolemic activity of synthetic and natural tocotrienols. J Med Chem 35:3595–3606

Pereira RI, Leitner JW, Erickson C et al (2008) Pioglitazone acutely stimulates adiponectin secretion from mouse and human adipocytes via activation of the phosphatidylinositol 3′-kinase. Life Sci 83:638–643

Perez-Ternero C, Herrera MD, Laufs U et al (2017) Food supplementation with rice bran enzymatic extract prevents vascular apoptosis and atherogenesis in ApoE−/− mice. Eur J Nutr 56:225–236

Phutthaphadoong S, Yamada Y, Hirata A et al (2010) Chemopreventive effect of fermented brown rice and rice bran (FBRA) on the inflammation-related colorectal carcinogenesis in ApcMin/+ mice. Oncol Rep 23:53–59

Pikarsky E, Porat RM, Stein I et al (2004) NF-κB functions as a tumour promoter in inflammation-associated cancer. Nature 431:461–466
Polakis P (2012) Wnt signaling in cancer. Cold Spring Harb Perspect Biol 4:a008052
Qureshi AA, Sami SA, Khan FA (2002) Effects of stabilized rice bran, its soluble and fiber fractions on blood glucose levels and serum lipid parameters in humans with diabetes mellitus types I and II. J Nutr Biochem 13:175–187
Raggi P, Genest J, Giles JT et al (2018) Role of inflammation in the pathogenesis of atherosclerosis and therapeutic interventions. Atherosclerosis 276:98–108
Raucci R, Rusolo F, Sharma A et al (2013) Functional and structural features of adipokine family. Cytokine 61:1–14
Ren Z, Pae M, Dao MC et al (2010) Dietary supplementation with tocotrienols enhances immune function in C57BL/6 mice. J Nutr 140:1335–1341
Ricciardiello L, Bazzoli F, Fogliano V (2011) Phytochemicals and colorectal cancer prevention-myth or reality? Nat Rev Gastroenterol Hepatol 8:592–596
Roschek B Jr, Fink RC, Li D et al (2009) Pro-inflammatory enzymes, cyclooxygenase 1, cyclooxygenase 2, and 5-lipooxygenase, inhibited by stabilized rice bran extracts. J Med Food 12:615–623
Saji N, Francis N, Schwarz LJ et al (2019) Rice bran derived bioactive compounds modulate risk factors of cardiovascular disease and type 2 diabetes mellitus: an updated review. Nutrients 11:2736
Salar A, Faghih S, Pishdad GR (2016) Rice bran oil and canola oil improve blood lipids compared to sunflower oil in women with type 2 diabetes: a randomized, single-blind, controlled trial. J Clin Lipidol 10:299–305
Sayin SI, Wahlstrom A, Felin J et al (2013) Gut microbiota regulates bile acid metabolism by reducing the levels of taurobeta-muricholic acid, a naturally occurring FXR antagonist. Cell Metab 17:225–235
Schmidt CG, Goncalves LM, Prietto L et al (2014) Antioxidant activity and enzyme inhibition of phenolic acids from fermented rice bran with fungus Rizhopus oryzae. Food Chem 146:371–377
Schwartz HM, Gilchrist FMC (1975) Microbial interactions with the diet and the host animal. In: McDonald IW, Warner ACI (eds) Digestion and metabolism in ruminant. The University of New England Publishing Unit, NSW, pp 165–179
Semenza GL (2010) Defining the role of hypoxia-inducible factor 1 in cancer biology and therapeutics. Oncogene 29:625–634
Senaphan K, Sangartit W, Pakdeechote P et al (2018) Rice bran protein hydrolysates reduce arterial stiffening, vascular remodeling and oxidative stress in rats fed a high-carbohydrate and high-fat diet. Eur J Nutr 57:219–230
Sharma JN, Mohammed LA (2006) The role of leukotrienes in the pathophysiology of inflammatory disorders: is there a case for revisiting leukotrienes as therapeutic targets? Inflammopharmacology 14:10–16
Sheflin AM, Borresen EC, Wdowik MJ et al (2015) Pilot dietary intervention with heat-stabilized rice bran modulates stool microbiota and metabolites in healthy adults. Nutrients 7:1282–1300
Sheflin AM, Borresen EC, Kirkwood JS et al (2017) Dietary supplementation with rice bran or navy bean alters gut bacterial metabolism in colorectal cancer survivors. Mol Nutr Food Res 61:1500905
Shehzad A, Lee J, Lee YS (2015) Autocrine prostaglandin E2 signaling promotes promonocytic leukemia cell survival via COX-2 expression and MAPK pathway. BMB Rep 48:109
Shen J, Yang T, Xu Y et al (2018) δ-tocotrienol, isolated from rice bran, exerts an anti-inflammatory effect via MAPKs and PPARs signaling pathways in lipopolysaccharide-stimulated macrophages. Int J Mol Sci 19:3022
Shibata A, Kawakami Y, Kimura T et al (2016) A-tocopherol attenuates the triglyceride- and cholesterol-lowering effects of rice bran tocotrienol in rats fed a western diet. J Agric Food Chem 64:5361–5366

Shuid AN, Das S, Mohamed IN (2019) Therapeutic effect of Vitamin E in preventing bone loss: an evidence-based review. Int J Vitam Nutr Res 89:357–370
Si X, Shang W, Zhou Z et al (2018) Gamma-aminobutyric acid enriched rice bran diet attenuates insulin resistance and balances energy expenditure via modification of gut microbiota and short-chain fatty acids. J Agric Food Chem 66:881–890
Siddiqui S, Ahsan H, Khan MR et al (2013) Protective effects of tocotrienols against lipid-induced nephropathy in experimental type-2 diabetic rats by modulation in TGF-b expression. Toxicol Appl Pharmacol 273:314–324
Singh BJ, Yadav D, Vij S (2019) Soybean bioactive molecules: current trend and future prospective. In: Mérillon JM, Ramawat K (eds) Bioactive molecules in food. Reference series in phytochemistry. Springer, Cham, pp 267–294
Sivagami G, Karthikkumar V, Balasubramanian T et al (2012) The modulatory influence of p-methoxycinnamic acid, an active rice bran phenolic acid, against 1,2-dimethylhydrazine-induced lipid peroxidation, antioxidant status and aberrant crypt foci in rat colon carcinogenesis. Chem Biol Interact 196:11–22
Song BL, DeBose-Boyd RA (2006) Insig-dependent ubiquitination and degradation of 3-hydroxy-3-methylglutaryl coenzyme a reductase stimulated by delta- and gamma-tocotrienols. J Biol Chem 281:25054–25061
Spadaro PA, Naug HL, Du Toit EF et al (2015) A refined high carbohydrate diet is associated with changes in the serotonin pathway and visceral obesity. Genet Res 97:e23
Stephanie H, Rekha G, Ion C et al (2015) Rice bran extract compensates mitochondrial dysfunction in a cellular model of early Alzheimer's disease. J Alzheimers Dis 43:927–938
Sugamura K, Keaney JF Jr (2011) Reactive oxygen species in cardiovascular disease. Free Radic Biol Med 51:978–992
Suzuki YJ, Tsuchiya M, Wassall SR et al (1993) Structural and dynamic membrane properties of alpha-tocopherol and alpha-tocotrienol: implication to the molecular mechanism of their antioxidant potency. Biochemist 32:10692–10699
Tan BL, Norhaizan ME (2017) Scientific evidence of rice by-products for cancer prevention: chemopreventive properties of waste products from rice milling on carcinogenesis *in vitro* and *in vivo*. Biomed Res Int 2017:9017902. 18p
Tan BL, Norhaizan ME, Suhaniza HJ et al (2013) Antioxidant properties and antiproliferative effect of brewers' rice extract (*temukut*) on selected cancer cell lines. Int Food Res J 20:2117–2124
Tan BL, Esa NM, Rahman HS et al (2014) Brewers' rice induces apoptosis in azoxymethane-induced colon carcinogenesis in rats via suppression of cell proliferation and the Wnt signaling pathway. BMC Complement Altern Med 14:304
Tan BL, Norhaizan ME, Huynh K et al (2015a) Water extract of brewers' rice induces apoptosis in human colorectal cancer cells via activation of caspase-3 and caspase-8 and downregulates the Wnt/β-catenin downstream signaling pathway in brewers' rice-treated rats with azoxymethane-induced colon carcinogenesis. BMC Complement Altern Med 15:205
Tan BL, Norhaizan ME, Huynh K et al (2015b) Brewers' rice modulates oxidative stress in azoxymethane-mediated colon carcinogenesis in rats. World J Gastroenterol 21:8826–8835
Tan BL, Norhaizan ME, Liew W-P-P (2018) Nutrients and oxidative stress: friend or foe? Oxidative Med Cell Longev 2018:9719584. 24p
Tetsu O, McCormick F (1999) β-catenin regulates expression of cyclin D1 in colon carcinoma cells. Nature 398:422–426
The GBD 2015 Obesity Collaborators (2017) Health effects of overweight and obesity in 195 countries over 25 years. N Engl J Med 377:13–27
Tsuchida A, Yamauchi T, Takekawa S et al (2005) Peroxisome proliferator-activated receptor (ppar)alpha activation increases adiponectin receptors and reduces obesity-related inflammation in adipose tissue: comparison of activation of pparalpha, ppargamma, and their combination. Diabetes 54:3358–3370
Tsuda H, Ohshima Y, Nomoto H et al (2004) Cancer prevention by natural compounds. Drug Metab Pharmacokinet 19:245–263

Upanan S, Yodkeeree S, Thippraphan P et al (2019) The proanthocyanidin-rich fraction obtained from red rice germ and bran extract induces HepG2 hepatocellular carcinoma cell apoptosis. Molecules 24:813

van der Woude CJ, Kleibeuker JH, Jansen PLM et al (2004) Chronic inflammation, apoptosis and (pre-)malignant lesions in the gastro-intestinal tract. Apoptosis 9:123–130

Vesely MD, Kershaw MH, Schreiber RD et al (2011) Natural innate and adaptive immunity to cancer. Annu Rev Immunol 29:235–271

Viana Abranches M, Esteves de Oliveira FC, Bressan J (2011) Peroxisome proliferator-activated receptor: effects on nutritional homeostasis, obesity and diabetes mellitus. Nutr Hosp 26:271–279

Vicennati V, Vottero A, Friedman C et al (2002) Hormonal regulation of interleukin-6 production in human adipocytes. Int J Obes Relat Metab Disord 26:905–911

Victor VM, Rocha M, Sola E et al (2009) Oxidative stress, endothelial dysfunction and atherosclerosis. Curr Pharm Des 15:2988–3002

Vissers MN, Zock PL, Meijer GW et al (2000) Effect of plant sterols from rice bran oil and triterpene alcohols from sheanut oil on serum lipoprotein concentrations in humans. Am J Clin Nutr 72:1510–1515

Wabitsch M, Funcke JB, von Schnurbein J et al (2015) Severe early-onset obesity due to bioinactive leptin caused by a p. N103K mutation in the leptin gene. J Clin Endocrinol Metab 100:3227–3230

Wang O, Liu J, Cheng Q et al (2015) Effects of ferulic acid and γ-oryzanol on high-fat and high-fructose diet-induced metabolic syndrome in rats. PLoS One 10:e0118135

Wang T, Gong X, Jiang R et al (2016) Ferulic acid inhibits proliferation and promotes apoptosis via blockage of PI3K/Akt pathway in osteosarcoma cell. Am J Transl Res 8:968

Whitehouse MW, Rainsford KD (2006) Lipoxygenase inhibition: the neglected frontier for regulating chronic inflammation and pain. Inflammopharmacology 14:99–102

Wolff SP, Dean RT (1987) Glucose autoxidation and protein modification. The potential role of 'autoxidative glycosylation' in diabetes. Biochem J 245:243–250

Wolff SP, Jiang ZY, Hunt JV (1991) Protein glycation and oxidative stress in diabetes mellitus and ageing. Free Radic Biol Med 10:339–352

World Health Organization (2019a) Cardiovascular disease. https://www.who.int/cardiovascular_diseases/en/. Accessed 18 June 2019

World Health Organization (2019b) Diabetes. https://www.who.int/health-topics/diabetes. Accessed 29 Dec 2019

World Health Organization (2019c) Dementia. https://www.who.int/news-room/fact-sheets/detail/dementia. Accessed 29 Dec 2019

World Health Organization (2019d) Chronic rheumatic conditions. https://www.who.int/chp/topics/rheumatic/en/. Accessed 30 Dec 2019

Xu W, Du M, Zhao Y et al (2012) γ-Tocotrienol inhibits cell viability through suppression of β-catenin/Tcf signaling in human colon carcinoma HT-29 cells. J Nutr Biochem 23:800–807

Yadav NV, Sadashivaiah, Ramaiyan B et al (2016) Sesame oil and rice bran oil ameliorates adjuvant-induced arthritis in rats: distinguishing the role of minor components and fatty acids. Lipids 51:1385–1395

Yamagishi SI, Edelstein D, Du XL et al (2001) Leptin induces mitochondrial superoxide production and monocyte chemoattractant protein-1 expression in aortic endothelial cells by increasing fatty acid oxidation via protein kinase A. J Biol Chem 276:25096–25100

Yerra VG, Kalvala AK, Sherkhane B et al (2018) Adenosine monophosphate-activated protein kinase modulation by berberine attenuates mitochondrial deficits and redox imbalance in experimental diabetic neuropathy. Neuropharmacology 131:256–270

Yousefian A, Leighton A, Fox K et al (2011) Understanding the rural food environment—perspectives of low-income parents. Rural Remote Health 11:1–11

Zgheib C, Hodges MM, Hu J et al (2017) Long non-coding RNA Lethe regulates hyperglycemia-induced reactive oxygen species production in macrophages. PLoS One 12:e0177453

Zha L, He L, Liang Y et al (2018) TNF-α contributes to postmenopausal osteoporosis by synergistically promoting RANKL-induced osteoclast formation. Biomed Pharmacother 102:369–374

Zhang JS, Li DM, He N et al (2011) A paraptosis-like cell death induced by δ-tocotrienol in human colon carcinoma SW620 cells is associated with the suppression of the Wnt signaling pathway. Toxicology 285:8–17

Chapter 6
Application in Food Products

Abstract Because the rice industry will remain sustainable for a long time, and thus the rice by-products production will remain high. Rice by-products are agriculture by-products produced during the rice milling process. They contain a variety of bioactive compounds including γ-oryzanol, phenolics, γ-aminobutyric acid (GABA), phytic acid, and dietary fibers. In view the therapeutic potential of rice by-products, its addition in food can contribute to the development of functional foods or value-added foods that currently are in high demand. The functional and nutritional properties of rice bran are well suited for the development of various foods such as pancakes, bread, and cookies. The addition of rice by-products in food industries has been demonstrated to improve the nutritional quality of processed food. In this regard, by-products produced from rice milling processes may enhance the economy among the rice producing nations. This chapter describes the potential application of rice by-products in various foods.

Keywords Coating · Color · Flavor · Foaming · Solubility · Stabilizer

Nutritious food is regarded as the primary source of well-being for consumers, mainly due to increased awareness of the association between well-being and nutritious food. Indeed, the product with good nutritional image, reasonable price, and good taste is in great need in recent decades (Sharif et al. 2014). Nonetheless, producing these food products is challenging because high nutritious foods such as high fiber content may lead to negative impacts on food quality and consumer acceptance. In this regard, food manufacturers are looking forward to the production of functional foods that are able to prevent diseases and enhance the health of individuals (Elisa et al. 2018). Table 6.1 shows the potential application of rice by-products in the food industry. Rice bran is rich in dietary fibers and has been utilized as a food additive to increase the fiber content in different foods (Martillanes et al. 2020). Stabilized rice bran or its derived compounds have been utilized as supplement in many food matrices, for instance, ground beef (Shih and Daigle 2003), milk powder (Nanuna et al. 2000), tuna oil (Chotimarkorn et al. 2008), beverages (Faccin et al. 2009), pizza (de Delahaye et al. 2005), breads (Hu et al. 2009), and cookies (Younas et al. 2011), owing to its high oil and water binding capacities that can maintain soft mouthfeel and reduce the moisture loss (Chia et al. 2015). Defatted

B. L. Tan, M. E. Norhaizan, *Rice By-products: Phytochemicals and Food Products Application*, https://doi.org/10.1007/978-3-030-46153-9_6

Table 6.1 Potential application of rice by-products in the food industry

Rice by-products	Purpose of addition	Products	References
Stabilized rice bran	High oil and water binding capacities	Ground beef	Shih and Daigle (2003)
		Milk powder	Nanuna et al. (2000)
		Tuna oil	Chotimarkorn et al. (2008)
		Beverages	Faccin et al. (2009)
		Pizza	de Delahaye et al. (2005)
		Breads	Hu et al. (2009)
		Cookies	Younas et al. (2011)
Rice bran	Improved solubility, protein extractability, flavor, and color	Baked goods and snacks	Sarkar and Bhattacharyya (1991)
	↑ Levels of protein	Cookies	Younas et al. (2011)
	↓ Cake volumes and ↑ weight and density	Sponge cake	Majzoobi et al. (2013)
	Improved the nutritional values such as fiber, γ-oryzanol, lutein, and antioxidant capacity	Cookies	de Souza et al. (2019)
	Enhanced the bioactive components and antioxidant activity	Bread	Irakli et al. (2015)
	Improved textural properties and dough rheological	Barbary bread	Milani et al. (2009)
Rice bran extract	↓ Peroxide production rate	Cookies	Bhanger et al. (2008)
	↑ Levels of antioxidants, total phenolics, and anthocyanins	Sausages	Loypimai et al. (2017)
Black glutinous rice bran	↑ Phytochemicals and provides good stability	Yogurt	Nontasan et al. (2012)
Rice bran protein	Improved the quality and properties of rice dough	Gluten-free rice noodles	Kim et al. (2014)
Infrared stabilized rice bran	↑ Levels of B vitamins, particularly niacin, and minerals (phosphorus, potassium, iron, and zinc)	Pan breads	Tuncel et al. (2014)
Rice bran wax	Protective layer	Vegetables and fruits	Dhall (2013)
	Better physicochemical properties, for example, juice and pH, total acidity, and firmness	Fruits and vegetables	Jutamongkon et al. (2011); Zhang et al. (2016)
	↓ Weight loss of pepper	Sweet pepper	Jutamongkon et al. (2011)

(continued)

Table 6.1 (continued)

Rice by-products	Purpose of addition	Products	References
	Edible coating	Candy such as gum and chocolate	Sabale et al. (2007); Dhall (2013)
Methanolic extracts of rice husk and rice bran	Enhanced stability	High-fat food products	Shih and Daigle (2003)
RBO	Better fat and water absorption properties	Shortening replacer	Sharif et al. (2005)
	Improved baking quality	Bread	Kaur et al. (2012)
	Improved baking quality	Cookies	Van Toan and Thanh Van (2019)
	Produces better quality in terms of organoleptic and baking properties	Bread	Orthoefer (2005)
	Improved organoleptic quality	Muffin	Kaur et al. (2014)
Broken rice	Used as an alternative for producing gluten-free products	Rice flour	Quiñones et al. (2015)
	As food additive to improve the satiety and bowel movements in pet	Pet food	De Godoy et al. (2013)
	To form a film (rice paper)	Traditional spring roll	Nagano et al. (2000)
Brewers' rice	Provides mouthfeel, color, aroma, and flavor	Beer	Marconi et al. (2017)
Rice straw	Selectively promoting the activity and/or proliferation of bacteria in the colon	Xylooligosaccharides	Chapla et al. (2013)

RBO rice bran oil

rice bran can be processed into flour, protein concentrates or in the forms of RBO (Prakash and Ramaswamy 1996). Figure 6.1 shows the physicochemical and functional properties of rice bran.

Rice bran exerts nutty flavor similar to peanut oil and thus is essential for baked goods and snacks (Sarkar and Bhattacharyya 1991). Indeed, the food quality is enhanced with the addition of rice bran, particularly solubility, protein extractability, flavor, and color, as well as other properties, for example foaming and emulsifying capacity, fat and water absorption (Madkour et al. 2018). The rice bran not only provided physiological benefits but also enhanced the stabilizing, thickening, gelling, and texture to certain foods (Sharma 1981).

Food color is one of the crucial factors for enhancing the appearance of food products and attracting consumer interest (Peanparkdee and Iwamoto 2019). Pigmented rice is a potential plant containing a variety of pigmented compounds including anthocyanins. This compound is not only exerting natural color but also

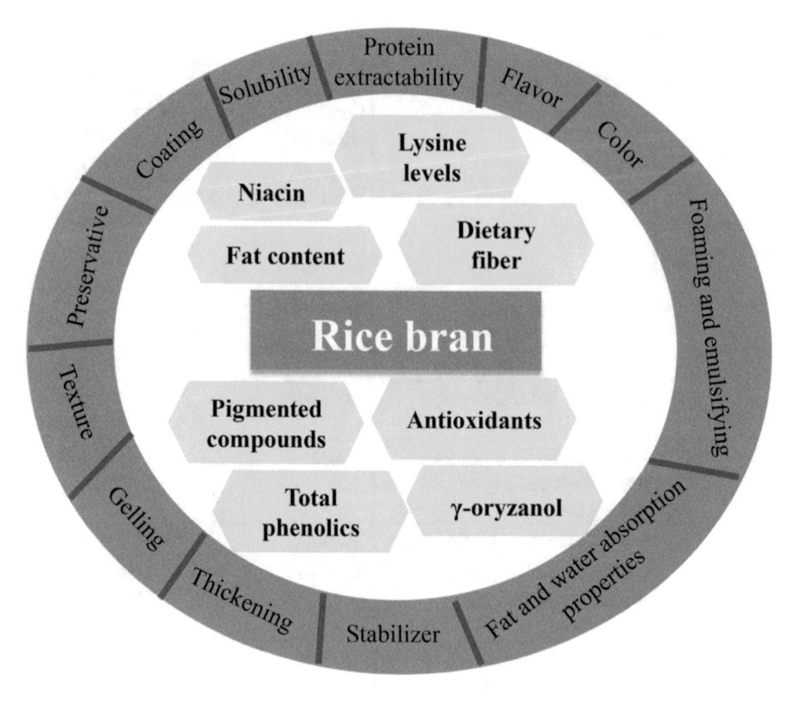

Fig. 6.1 The physicochemical and functional properties of rice bran

providing beneficial effects to human health (Peanparkdee and Iwamoto 2019). Many experiments have been conducted to assess the efficacy of by-products from pigmented rice as a natural colorant in food products. The data revealed that color of flavored yogurt was improved using the black glutinous rice extract containing γ-oryzanol, phenolic compounds, and anthocyanins. Furthermore, colorant powder from black glutinous rice bran increases the phytochemicals and provides good stability to the yogurt (Nontasan et al. 2012).

A study by Loypimai et al. (2017) evaluated the natural colorant from black glutinous rice bran. From the study reviewed, Loypimai et al. (2017) found that sausages produced from bran extract powder contain high amount of antioxidants, total phenolic content, and anthocyanins compared to the sausage containing 120 ppm or 0% of nitrite. The data further revealed that the application of bran extract powder showed better overall acceptance score and retarded lipid oxidation compared to the sausages containing 120 ppm or 0% of nitrite (Loypimai et al. 2017). Nitrite/nitrate is commonly used to suppress the growth of pathogenic bacteria including *Clostridium botulinum* and *Listeria* spp. Nitrite/nitrate can form nitrosamine when binding with the secondary amine in the stomach, which is carcinogenic to humans. Evidence from this study suggests that bran extract could be partially replaced nitrite in the fermented sausage product.

Foods containing high-fat content can cause rancidity due to lipid oxidation, which decreases the safety and nutritional value of food products and produces off-flavors (de Jesus and Gomes 2019). Historically, synthetic antioxidants such as butylated hydroxytoluene (BHT), propyl gallate, tert-butylhydroquinone (TBHQ), and butylated hydroxyanisole (BHA) are commonly used to prevent food spoilage and lipid oxidation and inhibit the formation of free radicals (Shih and Daigle 2003). Bioactive constituents from rice by-products have been suggested as one of the effective food stabilizers and preservatives, and thereby facilitate the inhibition of biologically harmful oxidation reactions. Therefore, these bioactive components can be applied as an alternative approach to minimize or avoid the use of synthetic food additives. Shih and Daigle (2003) further demonstrated that methanolic extracts of rice husk and rice bran could be used as a preservative agent to enhance the stability of foods high in fat. Research evidence has demonstrated that rice bran and rice husk extracts can effectively inhibit the lipid oxidation in ground beef, in which the efficiency of rice husk extract is similar to BHT. Due to their antioxidants, extracts from rice by-products are more likely to prolong the storage stability and inhibit lipid oxidation of beef products (Shih and Daigle 2003). Furthermore, the oxidative stability of rice bran extracts in cookies has also been evaluated over a period of one year at ambient storage conditions, without light (Bhanger et al. 2008). Bhanger et al. (2008) demonstrated that the addition of α-tocopherol, BHT, or rice bran extract in cookies reduced the rate of peroxide production compared to the control sample, indicating the high capability of rice bran extracts in the stabilization of cookies. This finding suggests that rice bran could be potentially used to extend the shelf-life of cookies.

The cookies containing more than 15% of defatted black rice bran showed the lowest overall acceptability (Elgammal et al. 2018), suggesting that high fiber content in defatted black rice bran compared to the wheat flour. Increasing concentration of rice bran produces darker cookies (Sudha et al. 2007). Darkening of cookies could be attributed to the Millard reaction between amino acids and sugars as well as sugar caramelization (Alobo 2001). Baked products containing rice bran flour, for example, pastries, muffins, and cookies have demonstrated a better nutritional value compared to wheat flour (Sarkar and Bhattacharyya 1991). Addition of 10, 15, and 20% of rice bran into wheat flour increased the protein levels in baked cookies by 0.93, 1.5, and 3.43%, respectively compared to cookies containing 100% wheat flour (Younas et al. 2011). Indeed, a small portion of fat in rice bran can serve as a carrier for flavor (Lavanya et al. 2017). Coating stabilized rice bran fibers on the chicken can reduce the fat absorption during frying (Hammond 1994). This finding implies that rice bran provides a better coating for food and requires less amount of oil to cook, and thus creating fewer polymers, which is associated with better flavor in food (Zareei et al. 2017).

Apart from that, the rice bran protein is gaining a lot of interest in the industrial food application (Phongthai et al. 2017). Rice bran protein is potential as nutritional supplements and functional food ingredients (Fabian and Ju 2011). Hydrolysis of protein with proteases could lead to the formation of peptide sequences; exert many antioxidative and functional properties (Hwang et al. 2010). Water solubility is one of the main characteristics of proteins since it can affect other properties such as gel-forming, foaming, and emulsion ability (Yeom et al. 2010). The elevation of protein solubility could be due to the unfolding of the protein molecule and smaller molecular peptides (Stefani and Dobson 2003). At extremely alkaline and acidic pHs, proteins carry net negative and positive charges, respectively, and thus ionic hydration and electrostatic repulsion enhanced the solubilization of the protein (Yeom et al. 2010). Water and oil absorption capacity may also affect the main properties of rice bran protein in food systems (Chandi and Sogi 2007). The underlying mechanism of oil absorption capacity could be due to the hydrophobicity of the material and the combination of physical entrapment of oil (Binks and Lumsdon 2000). Exposure to hydrophobic groups and unfolding of protein structure allows the physical entrapment of oil (Vácha et al. 2011). The low hydrophobicity of rice protein does not facilitate the interaction between oil and proteins and thus reduced oil absorption capacity (Zhang et al. 2012). High oil absorption capacity is essential in the formulation of cake batters and sausages, while high water absorption of protein may decrease the moisture loss in bakery products. Foaming ability is another basic functional property of proteins. The elevation of protein solubility through proteolysis may improve foaming capacity. However, extensive hydrolysis may decrease foaming due to an excessive charge that hinders the formation of stable foam (Zhang et al. 2012).

Furthermore, rice bran protein can form an excellent base for high sugar foods such as confections, frozen desserts, and cake batters (Ghorbani-HasanSaraei et al. 2019). Kim et al. (2014) used rice bran protein isolate to improve the properties of rice dough and quality of gluten-free rice noodles. The data showed that rice bran protein isolate (10%) improved the microstructure of the noodles, cooking quality, and nutritional quality (Kim et al. 2014). The functional properties of sponge cake and batter were evaluated by different particle sizes (210, 125, and 53 μm) and concentrations (0, 5, 10, 15, and 20%) of rice bran (Majzoobi et al. 2013). Batter and sponge cake containing larger particle size and higher quantity of bran was associated with a reduction in cake volumes and increased weight and density of cakes. The addition of 10% rice bran with a particle size of 125 μm showed the most desirable cake quality (Majzoobi et al. 2013).

The addition of rice bran into the wheat flour not only enhanced the lysine and protein contents, it also increased the components of dietary fiber in cookies and bread (Ajmal et al. 2006; Jisha et al. 2010). A recent study has developed five different formulations of cookies containing 25, 50, 75, and 100% rice bran. The sensory analysis showed that incorporation of rice bran into the cookies showed higher acceptance levels compared to the control cookies, with the preference for the cookies containing 75% of rice bran (de Souza et al. 2019). The data further revealed that incorporation of cookies with rice bran significantly improved the nutritional values such as antioxidant capacity, lutein, γ-oryzanol, and fiber contents compared to the control cookies ($p < 0.05$) (de Souza et al. 2019). This finding implies that rice bran could be used as an alternative to enhance the nutritional content in cookies.

Irakli et al. (2015) studied the effects of substitution wheat bread with rice bran on antioxidant properties, bioactive components, and quality attributes. The study has shown that replacing wheat bread with rice bran at ratio 1:4 (w/w) enhanced the bioactive components and antioxidant activity of bread, without affecting the sensory attributes and overall bread quality (Irakli et al. 2015). Rice bran fiber exerts a high amount of functional fats and proteins as well as trace minerals, vitamins, and antioxidants (Raghav et al. 2016). The recent study found that superfine rice bran insoluble dietary fiber powder showed higher antioxidant properties and phenolic bioaccessibility, higher extractability in bound and free phenolics compared to the coarse counterpart (Zhao et al. 2018). This finding suggests that superfine grinding can enhance the functional properties of rice bran insoluble dietary fiber and could be applied in functional foods (Zhao et al. 2018). Replacement of wheat flour with infrared stabilized rice bran at concentrations of 2.5, 5.0, and 10.0% in whole grain wheat, wheat bran, and white wheat bread was shown to increase the levels of B vitamins, particularly niacin (Tuncel et al. 2014). The addition of infrared stabilized rice bran in the pan bread also significantly increased the phosphorus, potassium, iron, and zinc levels (Tuncel et al. 2014). The textural properties and dough rheological of Barbary bread were determined by the addition of different concentrations of rice bran flour (9, 6, 3, and 0%) to wheat flour (88 and 82% extraction rate) (Milani et al. 2009). The data from the sensory analysis revealed that blending rice bran flour (6%) along with the wheat flour (82% extraction rate) showed the highest acceptable sensory characteristics (Milani et al. 2009). Collectively, rice bran can be applied as inexpensive compounds and thereby provide optimal dietary and nutritional supplements for overall health maintenance. Nonetheless, further studies are warranted to provide detailed information and recommendations to individual consumer requirements.

From a marketing perspective, RBO is the most common rice bran-derived component. RBO is produced from the outer layer of rice and usually is commercialized in the form of refined oil (Van Toan and Thanh Van 2019). The oil has been recognized to contain a better flavor and taste to food products, more stable at high temperatures, longer shelf-life, more micronutrients, rich in PUFA, and high free fatty acids (Van Toan and Thanh Van 2019) (Fig. 6.2). RBO is hypoallergenic and thus can be utilized as alternative cooking oil for those who are allergic to conventional oil (Nayik et al. 2015). In addition, RBO is suitable for stir-frying and deep-frying due to its high stability at high cooking temperatures (Sharma 2002). RBO has been used as a substitution in several food products including salad dressing and mayonnaise (Kaur et al. 2014). The addition of RBO into the roast beef showed higher oxidative stability during storage compared to the roast beef without RBO (Kim et al. 2000). RBO is also utilized in baked foods as a shortening replacer because baked products require an emulsifier and fat due to their fat and water absorption properties (Sharif et al. 2005). Replacement shortening with 50% RBO showed an improvement in the baking quality of bread (Kaur et al. 2012). In line with this, the study also found that substitution vegetable shortening with Vietnamese RBO at the ratio of 50% improved the quality of cookies (Van Toan and Thanh Van 2019). The sensory evaluation analysis further revealed that cookies containing 50–75% of RBO showed the highest acceptable sensory characteristics (Van Toan and Thanh Van 2019). In addition, the replacement of conventional shortening with 50% of RBO produces better quality bread in terms of organoleptic and baking properties (Orthoefer 2005). Besides cookies and bread, RBO is also potential to produce muffin. The data showed that

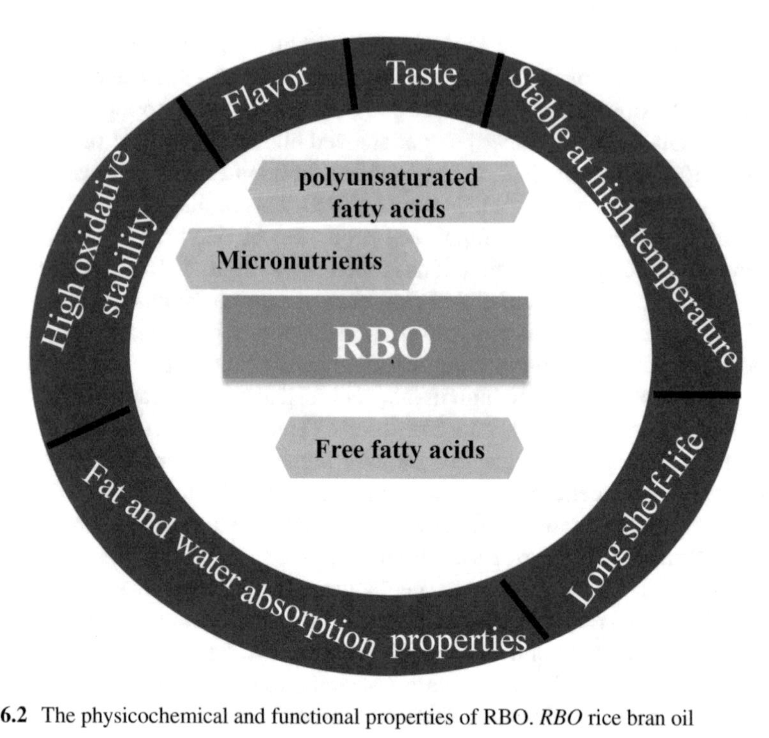

Fig. 6.2 The physicochemical and functional properties of RBO. *RBO* rice bran oil

organoleptic quality of muffin prepared by replacing bakery shortening with 50 or 75% of RBO varied significantly which is desirable from that of control (Kaur et al. 2014). Besides its potential in the food industry, previous findings suggested that oil in rice bran can facilitate the binding of animal feed, and thus decreases feed waste during feeding (Bodie et al. 2019). This finding implies that RBO is not only used in food manufacturers but can also be applied in the poultry industry, demonstrating the enormous potential of RBO.

Besides rice bran and RBO, rice bran wax can also provide an inexpensive alternative coating for vegetables and fruits that served as a protective layer (Dhall 2013). A study showed that adding rice bran wax (10% as a covering) into sweet pepper can reduce weight loss of pepper after 7 days (Jutamongkon et al. 2011). Coating fruits and vegetables with rice bran wax showed better physicochemical properties, for example, juice and pH, total acidity, and firmness (Jutamongkon et al. 2011; Zhang et al. 2016) (Fig. 6.3). Rice bran wax has also been used as an edible coating to candy, for example, gum and chocolate (Sabale et al. 2007; Dhall 2013). In a study by Dassanayake et al. (2009) focusing on crystal morphology and thermal behavior in rice bran wax, carnauba wax, and candelilla wax, rice bran wax showed a better ability to structure oils, with a minimum level at 0.5% of rice bran being able to form a gelation compared to carnauba wax (4%) and candelilla wax (2%). Despite rice bran wax is potential to be used as a food additive; the use of rice bran wax is currently restricted on its use. Hence, further studies are required to potentially overcome the current restrictions and their utilization.

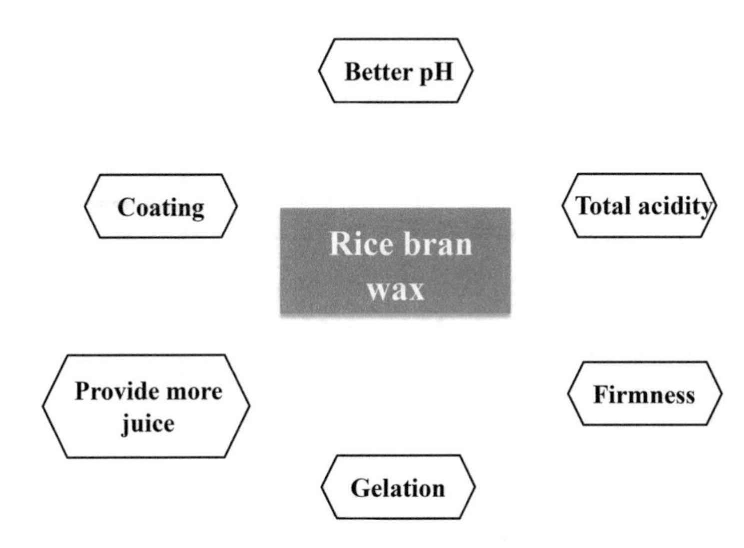

Fig. 6.3 The functional properties of rice bran wax

Broken rice can be processed into flour and used as a food additive because of its nutritional values (Kim et al. 2012). Gluten is one of the contributors to celiac disease (Hartmann et al. 2006). Several foods such as pasta, cereals, snacks, and baked goods contained high amounts of gluten. Rice flour is a gluten-free product and thus is potential to be used as an alternative for producing gluten-free products (Quiñones et al. 2015). In addition, rice flour can also be utilized to process into puddings and baby food. The food industry prefers rice flour rather than other flours because it can decrease the risk in people with sensitivities (Gujaral and Rosell 2004). In this regard, it has become more economically justifiable to use broken rice flour for such applications (Qian and Zhang 2013).

Broken rice has been widely used as pet food in the United States (Buff et al. 2014). It is a popular additive as it can facilitate the bowel movements and improve the satiety in pets due to its high fiber contents (De Godoy et al. 2013). The broken rice has also been consumed in humans. For instance, Banh Da Nem, one of the broken rice dishes that are popular in Vietnam, in which the broken rice are broken down and combined with water to form a film (rice paper) which is used to form a traditional spring roll (Nagano et al. 2000). As described in Chap. 3, brewers' rice contained a mixture of rice germ and rice bran with broken kernels (Nordin et al. 2014). It is commonly utilized as beer brewing ingredients. Brewers' rice provides mouthfeel, color, aroma, and flavor to beer (Marconi et al. 2017) (Fig. 6.4). It can also used as a raw material that acts as a substrate for the yeast to ferment and ultimately produce alcohol. Compared to barley malt, brewers' rice is the grain of choice for beer industries because it is more economical. Nevertheless, beers containing rice provides light color, very neutral, and dry flavor of beer (Marconi et al. 2017).

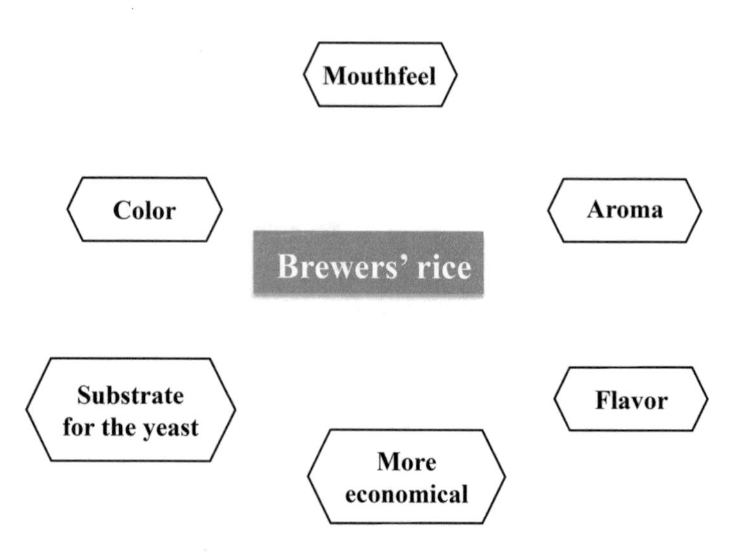

Fig. 6.4 The functional properties of brewers' rice

Notably, the enzymatic hydrolysis of xylan extracted from agro-residues, for instance, corn straw, wheat straw, and rice straw can be used to produce xylooligosaccharides (XOS) (Otieno and Ahring 2012; Chapla et al. 2013; Surek and Buyukkileci 2017). XOS is a prebiotic or nondigestible food ingredient that beneficially influences the host by selectively promoting the activity and/or proliferation of bacteria in the colon, and thereby promoting host health (Gibson and Roberfroid 1995). Chapla et al. (2013) exploring the enzymatic production of XOS from xylan of rice straw using β-xylosidase free xylanase. The prebiotic potential of XOS was verified using *in vitro* fermentation of XOS using *Bifidobacterium* spp., which implied that XOS has probiotic nature. This result indicates that XOS can be utilized in acidic and heat-processed food industries.

References

Ajmal M, Butt MS, Sharif K et al (2006) Preparation of fiber and mineral enriched pan bread by using defatted rice bran. Int J Food Prop 9:623–636

Alobo AP (2001) Effect of sesame seed flour on millet biscuit characteristics. Plant Food Hum Nutr 56:195–202

Bhanger MI, Iqbal S, Anwar F et al (2008) Antioxidant potential of rice bran extracts and its effects on stabilisation of cookies under ambient storage. Int J Food Sci Technol 43:779–786

Binks BP, Lumsdon SO (2000) Effects of oil type and aqueous phase composition on oil–water mixtures containing particles of intermediate hydrophobicity. Phys Chem Chem Phys 2:2959–2967

Bodie AR, Micciche AC, Atungulu GG et al (2019) Current trends of rice milling byproducts for agricultural applications and alternative food production systems. Front Sustain Food Syst 3:47

Buff PR, Carter RA, Bauer JE et al (2014) Natural pet food: a review of natural diets and their impact on canine and feline physiology. J Anim Sci 92:3781–3791

Chandi G, Sogi DS (2007) Functional properties of rice bran protein concentrates. J Food Eng 79:592–597
Chapla D, Dholakiya S, Madamwar D et al (2013) Characterization of purified fungal endoxylanase and its application for production of value added food ingredient from agroresidues. Food Bioprod Process 91:682–692
Chia SL, Boo HC, Muhamad K et al (2015) Effect of subcritical carbon dioxide extraction and bran stabilization methods on rice bran oil. J Am Oil Chem Soc 92:393–402
Chotimarkorn C, Benjakul S, Silalai N (2008) Antioxidative effects of rice bran extracts on refined tuna oil during storage. Food Res Int 41:616–622
Dassanayake LSK, Kodali DR, Ueno S et al (2009) Physical properties of rice bran wax in bulk and organogels. J Am Oil Chem Soc 86:1163–1174
de Delahaye EP, Jimenez P, Perez E (2005) Effect of enrichment with high content dietary fibre stabilized rice bran flour on chemical and functional properties of storage frozen pizzas. J Food Eng 68:1–7
De Godoy MR, Kerr KR, Fahey GC (2013) Alternative dietary fiber sources in companion animal nutrition. Nutrients 5:3099–3117
de Jesus DJF, Gomes DC (2019) Comparative analysis of rancidity development in vegetable oils during frying process. J Appl Pharm Sci 6:65–74
de Souza CB, Lima GPP, Vanz Borges C et al (2019) Development of a functional rice bran cookie rich in γ-oryzanol. J Food Meas Charact 13:1070–1077
Dhall RK (2013) Advances in edible coatings for fresh fruits and vegetables: a review. Crit Rev Food Sci Nutr 53:435–450
Elgammal RE, Rabie MA, El Bana MA et al (2018) Processing cookies from defatted thermal stabilized black rice bran. J Food Dairy Sci, 3rd Mansoura International Food Congress (MIFC), 1–5 Oct 2018
Elisa DT, Mariarosaria S, Debora S (2018) Consumer acceptance and consumption of functional foods. An attempt of comparison between Italy and Germany. Food Saf Manag 19:125–132
Fabian C, Ju Y-H (2011) A review on rice bran protein: its properties and extraction methods. Crit Rev Food Sci Nutr 51:816–827
Faccin GL, Vieira LN, Miotto LA et al (2009) Chemical, sensorial and rheological properties of a new organic rice bran beverage. Rice Sci 16:226–234
Ghorbani-HasanSaraei A, Rafe A, Shahidi S-A et al (2019) Microstructure and chemorheological behavior of whipped cream as affected by rice bran protein addition. Food Sci Nutr 7:875–881
Gibson GR, Roberfroid MB (1995) Dietary modulation of the human colonic microbiota: introducing the concept of prebiotics. J Nutr 125:1401–1412
Gujaral HS, Rosell CM (2004) Improvement of breadmaking quality of rice flour by glucose oxidase. Food Res Int 37:75–81
Hammond N (1994) Functional and nutritional characteristics of rice bran extracts. Cereal Foods World 39:752–754
Hartmann G, Koehler P, Wieser H (2006) Rapid degradation of gliadin peptides toxic for coeliac disease patients by proteases from germinating cereals. J Cereal Sci 44:368–371
Hu G, Huang S, Cao S et al (2009) Effect of enrichment with hemicellulose from rice bran on chemical and functional properties of bread. Food Chem 115:839–842
Hwang JY, Shyu YS, Wang YT et al (2010) Antioxidative properties of protein hydrolysate from defatted peanut kernels treated with esperase. LWT Food Sci Technol 43:285–290
Irakli M, Katsantonis D, Kleisiaris F (2015) Evaluation of quality attributes, nutraceutical components and antioxidant potential of wheat bread substituted with rice bran. J Cereal Sci 65:74–80
Jisha S, Padmaja G, Sajeev MS (2010) Nutritional and textural studies on dietary fiber-enriched muffins and biscuits from cassava-based composite flours. J Food Qual 33:79–99
Jutamongkon R, Praditdoung S, Vananuvat N (2011) Effect of rice bran waxing on fruit and vegetable storage. Kasetsart J Nat Sci 45:1115–1126
Kaur A, Jassal V, Thind SS et al (2012) Rice bran oil an alternate bakery shortening. J Food Sci Technol 49:110–114

Kaur A, Jassal V, Bhise SR (2014) Replacement of bakery shortening with rice bran oil in the preparation of muffins. Afr J Biochem Res 8:141–146

Kim JS, Godber JS, Prinaywiwatkul W (2000) Restructured beef roasts containing rice bran oil and fiber influences cholesterol oxidation and nutritional profile. J Muscle Foods 11:111–127

Kim HY, Hwang IG, Kim TM et al (2012) Chemical and functional components in different parts of rough rice (*Oryza sativa* L.) before and after germination. Food Chem 134:288–293

Kim Y, Kee JI, Lee S et al (2014) Quality improvement of rice noodle restructured with rice protein isolate and transglutaminase. Food Chem 145:409–416

Lavanya MN, Venkatachalapathy N, Manickavasagan A (2017) Physicochemical characteristics of rice bran. In: Manickavasagan A, Santhakumar C, Venkatachalapathy N (eds) Brown rice. Springer, Cham, pp 79–90

Loypimai P, Moongngarm A, Naksawat S (2017) Application of natural colorant from black rice bran for fermented Thai pork sausage-Sai Krok Isan. Int Food Res J 24:1529–1537

Madkour AH, Allam MH, Abdel Fattah AA et al (2018) Functional, rheological and sensory characteristics of defatted-hydrolyzed rice bran as fat replacers in prepared biscuit. J Agric Sci 26:1509–1519

Majzoobi M, Sharifi S, Imani B et al (2013) The effect of particle size and level of rice bran on the batter and sponge cake properties. J Agric Sci Technol 15:1175–1184

Marconi O, Sileoni V, Ceccaroni D et al (2017) Chapter 4: the use of rice in brewing. In: Marconi (ed) Advances in international rice research. International Rice Research Technology, London, pp 50–64

Martillanes S, Rocha-Pimienta J, Gil MV et al (2020) Antioxidant and antimicrobial evaluation of rice bran (*Oryza sativa* L.) extracts in a mayonnaise-type emulsion. Food Chem 308:125633

Milani E, Pourazarang H, Mortazavi S (2009) Effect of rice bran addition on dough rheology and textural properties of barbary bread. (in Persian). Iran J Food Sci Technol 6:23–31

Nagano H, Shoji Z, Tamura A et al (2000) Some characteristics of rice paper of Vietnamese traditional food (Vietnamese spring rolls). Food Sci Technol Res 6:102–105

Nanuna JN, McGregor JU, Godber JS (2000) Influence of high-oryzanol rice bran oil on the oxidative stability of whole milk powder. J Dairy Sci 83:2426–2431

Nayik GA, Majid I, Gull A et al (2015) Rice bran oil, the future edible oil of India: a mini review. J Rice Res 3:1–3

Nontasan S, Moongngarm A, Deeseenthum S (2012) Application of functional colorant prepared from black rice bran in yogurt. APCBEE Proc 2:62–67

Nordin NNAM, Karim R, Ghazali HM et al (2014) Effects of various stabilization techniques on the nutritional quality and antioxidant potential of brewer's rice. J Eng Sci Technol 9:347–363

Orthoefer FT (2005) Rice bran oil. In: Bailey's industrial oil and fat products, vol. 2, pp 465–489

Otieno DO, Ahring BK (2012) A thermochemical pretreatment process to produce xylooligosaccharides (XOS), arabinooligosaccharides (AOS) and mannooligosaccharides (MOS) from lignocellulosic biomasses. Bioresour Technol 112:285–292

Peanparkdee M, Iwamoto S (2019) Bioactive compounds from by-products of rice cultivation and rice processing: extraction and application in the food and pharmaceutical industries. Trends Food Sci Technology 86:109–117

Phongthai S, Homthawornchoo W, Rawdkuen S (2017) Preparation, properties and application of rice bran protein: a review. Int Food Res J 24:25–34

Prakash J, Ramaswamy HS (1996) Rice bran proteins: properties and food uses. Crit Rev Food Sci Nutr 36:537–552

Qian H, Zhang H (2013) Rice flour and related products. In: Bhandari B, Bansal N, Zhang M et al (eds) Handbook of food powders: processes and properties. Woodhead Publishing, Philadelphia, pp 553–575

Quiñones RS, Macachor CP, Quiñones HG (2015) Development of gluten-free composite flour blends. Trop Technol J 19:1–4

Raghav PK, Agarwal N, Sharma A (2016) Emerging health benefits of rice bran—a review. Int J Multidiscip Res Mod Educ 2:367–382

Sabale V, Sabale PM, Lakhotiya CL (2007) In vitro studies on rice bran oil wax as skin moisturizer. Indian J Pharm Sci 69:215–218
Sarkar S, Bhattacharyya DK (1991) Nutrition of rice bran oil in relation to its purification. J Am Oil Chem Soc 68:956–962
Sharif K, Butt MS, Anjum FM et al (2005) Improved quality of baked products by rice bran oil. Int J Food Saf 5:1–8
Sharif MK, Butt MS, Anjum FM et al (2014) Rice bran: a novel functional ingredient. Crit Rev Food Sci Nutr 54:807–816
Sharma SC (1981) Gums and hydrocolloids in oil-water emulsion. Food Technol 35:59–67
Sharma AR (2002) Edible rice bran oil consumer awareness programme. In: Rice bran oil promotion committee. Solvent extractors Association of India, Mumbai
Shih FF, Daigle KW (2003) Antioxidant properties of milled-rice co-products and their effects on lipid oxidation in ground beef. J Food Sci 68:2672–2675
Stefani M, Dobson CM (2003) Protein aggregation and aggregate toxicity: new insights into protein folding, misfolding diseases and biological evolution. J Mol Med 81:678–699
Sudha MLR, Vetrimani K, Leelavathi K (2007) Influence of fibre from different cereals on the rheological characteristics of wheat flour dough and on biscuit quality. Food Chem 100:1365–1370
Surek E, Buyukkileci AO (2017) Production of xylooligosaccharides by autohydrolysis of hazelnut (Corylus avellana L.) shell. Carbohydr Polym 174:565–571
Tuncel NB, Yilmaz N, Kocabiyik H et al (2014) The effect of infrared stabilized rice bran substitution on B vitamins, minerals and phytic acid content of pan breads: part II. J Cereal Sci 59:162–166
Vácha R, Rick SW, Jungwirth P et al (2011) The orientation and charge of water at the hydrophobic oil droplet–water interface. J Am Chem Soc 133:10204–10210
Van Toan N, Thanh Van NT (2019) Effects of Vietnamese rice bran oil as vegetable shortening substitution on the physical and sensory of baked cookies. Clin J Nutr Diet 2:1–9
Yeom HJ, Lee EH, Ha MS et al (2010) Production and physicochemical properties of rice bran protein isolates prepared with autoclaving and enzymatic hydrolysis. J Korean Soc Appl Biol Chem 53:62–70
Younas A, Bhatti MS, Ahmed A et al (2011) Effect of rice bran supplementation on cookie baking quality. Pak J Agric Sci 48:133–138
Zareei SA, Ameri F, Dorostkar F et al (2017) Rice husk ash as a partial replacement of cement in high strength concrete containing micro silica: evaluating durability and mechanical properties. Case Stud Constr Mater 7:73–81
Zhang HJ, Zhang H, Wang L et al (2012) Preparation and functional properties of rice bran proteins from heat-stabilized defatted rice bran. Food Res Int 47:359–363
Zhang L, Chen F, Zhang P et al (2016) Influence of rice bran wax coating on the physicochemical properties and pectin nanostructure of cherry tomatoes. Food Bioprocess Technol 10:349–357
Zhao G, Zhang R, Dong L et al (2018) Particle size of insoluble dietary fiber from rice bran affects its phenolic profile, bioaccessibility and functional properties. LWT Food Sci Technol 87:450–456

Chapter 7
Summary and Future Prospects

Abstract Rice is one of the universal cereal crops consumed by nearly half of the world population as their daily staple food. Rice milling process produces several by-products including rice bran, rice straw, rice germ, broken rice, rice husk, and brewers' rice. In particular, rice by-products are rich in bioactive constituents. Many studies have revealed that rice by-products can potentially promote body weight loss, improve insulin sensitivity and glucose control, decrease the risk of cancer, slow down the development of neurodegenerative disorders, improve lipid profiles, alleviate arthritis, and decrease the risk of osteoporosis. Consumer attitude towards nutritious foods is promising and thus the demands of functional foods are increased in the world market. Furthermore, the addition of food additive has become a major concern among consumers. In this regard, rice by-products have been applied as food preservatives and stabilizers, food ingredients for the development of value-added food products, and colorants, as well as utilized as alternative poultry production and fuel industry. Taken together, the utilization of rice by-products in various industries could improve efficiency and reduce waste and thereby decreasing environmental problems. Rice by-products seem a good dietary agent for treating metabolic ailments and providing nutrients.

Keywords Food enrichment · Neuroprotective agent · Pharmaceutical industry

Rice is a crucial energy source with a broad spectrum of functional properties. Rice grains need to undergo some rice processing procedures before being consumed as food. Rice milling is a crucial post-harvesting process that affects the nutritional quality. The germs and some of the endosperm such as powdery material and broken rice are separated during the rice milling process. The output of rice milling comprised of primary products, namely milled rice, and rice by-products including broken kernels, bran layer, germ, and husk (Ryan 2011). Rice by-products are good source of minerals, fatty acids, fibers, vitamins, and proteins. Recently, there has been an increase in the utilization of rice by-products, especially broken kernels and rice bran (Esa et al. 2013; Tan and Norhaizan 2017; Triratanasirichai et al. 2017). Previous studies have demonstrated the enormous functional role of rice by-products including broken rice, rice husk, rice straw, rice bran, and brewers' rice (Esa et al. 2013; Tan et al. 2014, 2015; Tan and Norhaizan 2017). Among all the rice

B. L. Tan, M. E. Norhaizan, *Rice By-products: Phytochemicals and Food Products Application*, https://doi.org/10.1007/978-3-030-46153-9_7

by-products, rice bran is the most extensively studied. Rice bran contains high level of fibers, minerals, vitamins, and bioactive constituents, and thus gaining a lot of interest in the treatment and prevention of chronic diseases (Dokkaew et al. 2019; Al-Okbi et al. 2020). Since phenolic components offer beneficial health benefits, this may partly explain a better nutritional value of rice by-products (Meselhy et al. 2019; Aalim et al. 2019). Compelling evidence from a preclinical study both in *in vivo* and *in vitro* models demonstrated that rice by-products exhibit encouraging findings in the alleviation of wide range diseases. Figures 7.1 and 7.2 summarize the mechanisms involved in rice by-products in relation to obesity and cancer, CVD and diabetes, respectively.

Health beneficial components in pigmented rice have shown many bioactivities such as antiallergic, hypoglycemic, anti-atherosclerosis, antioxidant, and antitumor (Deng et al. 2013). The mechanisms of rice by-products in neurodegenerative disease, osteoporosis, and arthritis are shown in Fig. 7.3. The underlying molecular mechanism that has been suggested to combat these diseases could be partly through potentiation of bioactive constituents such as essential oils, GABA, amino acids, γ-oryzanol, tocopherols and tocotrienols, sterols, phenolics, tannins, and flavones (Tan and Norhaizan 2017). The potential implication of rice by-products in relation to chronic diseases worth further elucidation in randomized clinical trials.

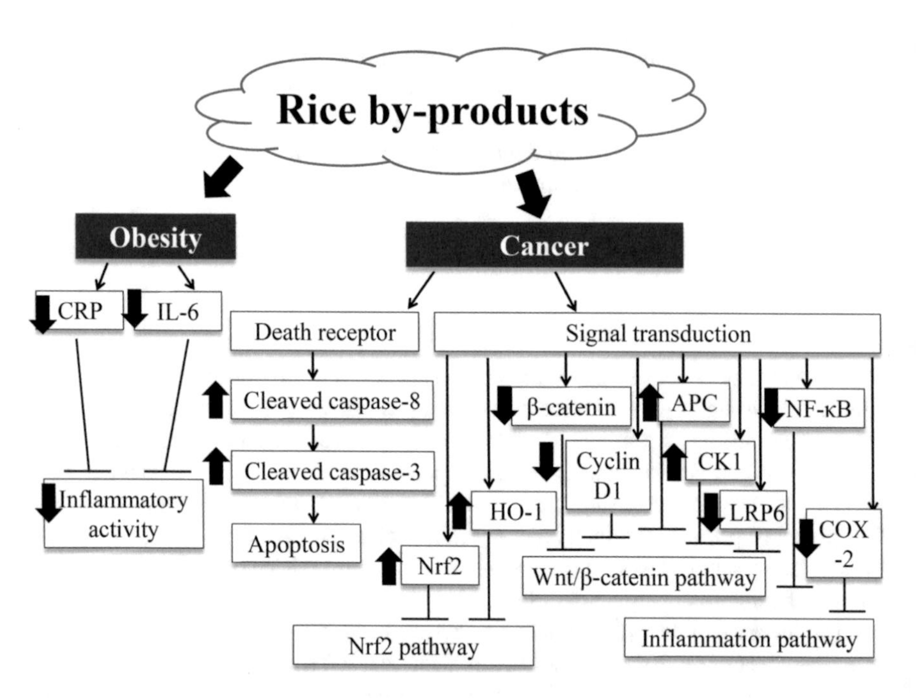

Fig. 7.1 The mechanisms involved in rice by-products in relation to obesity and cancer. *APC* adenomatous polyposis coli, *CK1* casein kinase 1, *COX-2* cyclooxygenase-2, *CRP* C-reactive protein, *HO-1* heme oxygenase-1, *IL-6* interleukin-6, *LRP6* low-density lipoprotein receptor-related protein 6, *NF-κB* nuclear factor-kappa B, *Nrf2* NF-E2-related factor 2

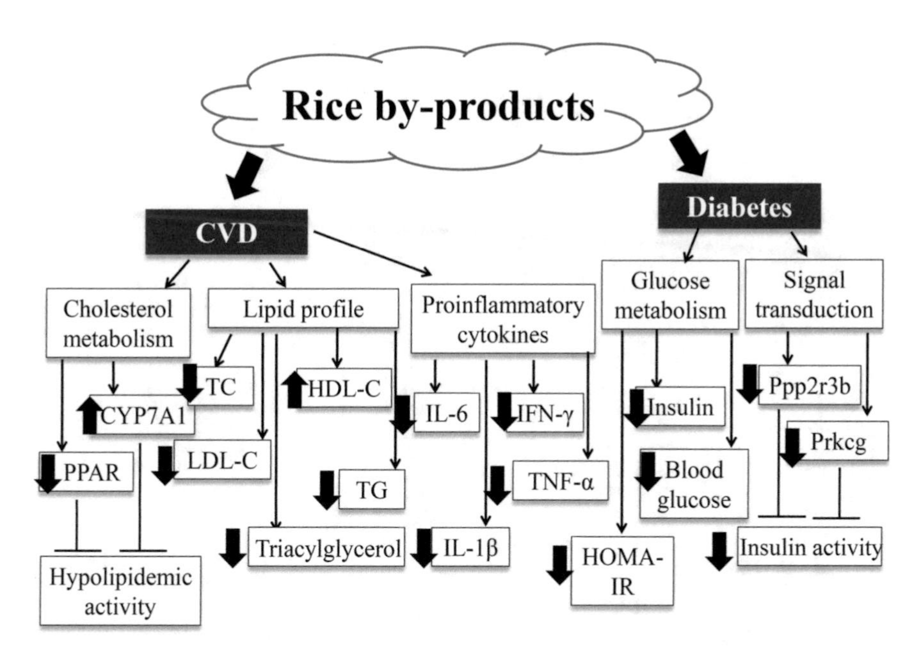

Fig. 7.2 The mechanisms involved in rice by-products in relation to CVD and diabetes. *CVD* cardiovascular disease, *HDL-C* high-density lipoprotein cholesterol, *HOMA-IR* homeostasis model assessment-insulin resistance, *IFN-γ* interferon-gamma, *IL-1β* interleukin-1beta, *IL-6* interleukin-6, *LDL-C* low-density lipoprotein cholesterol, *PPAR* peroxisome proliferator-activated receptor, *TC* total cholesterol, *TG* triglycerides, *TNF-α* tumor necrosis factor-alpha

Phytochemicals are well-recognized in maintaining the oxidative-antioxidant balance (Tan et al. 2018). It is known for its wide range of health-promoting potential on animals and humans (Gul et al. 2016; Panche et al. 2016). Hence, many processing treatments have been applied for the extraction of bioactive constituents from the rice by-products (Peanparkdee and Iwamoto 2019). Indeed, the polarity and type of solvents used for extracting the bioactive components depend on the solubility of the phytochemicals (Felhi et al. 2017). From an industry perspective, the primary factors to be considered are the toxicity of extractant solvents and extraction cost of bioactive constituents. Indeed, extraction technologies that are environmental friendly, effective, applicable, low cost, and simple are needed to expand the application of by-products from rice processing in developing countries. In order to reduce environmental pollution, greater attention should be paid to the application of these by-products in the food industries. Despite many extraction conditions and techniques that have been optimized for extracting the bioactive compounds in rice by-products, the applications require further elucidation.

Considering the bioactive components of rice by-products, it has extended to several fields, particularly in the pharmaceutical and food industries. In the food industry, it has been applied as a food stabilizer and preservative, colorants, and ingredients for the development of food products to increase shelf-life and enhance the texture, color, and nutritional value of products (Peanparkdee and Iwamoto 2019). In addition, the

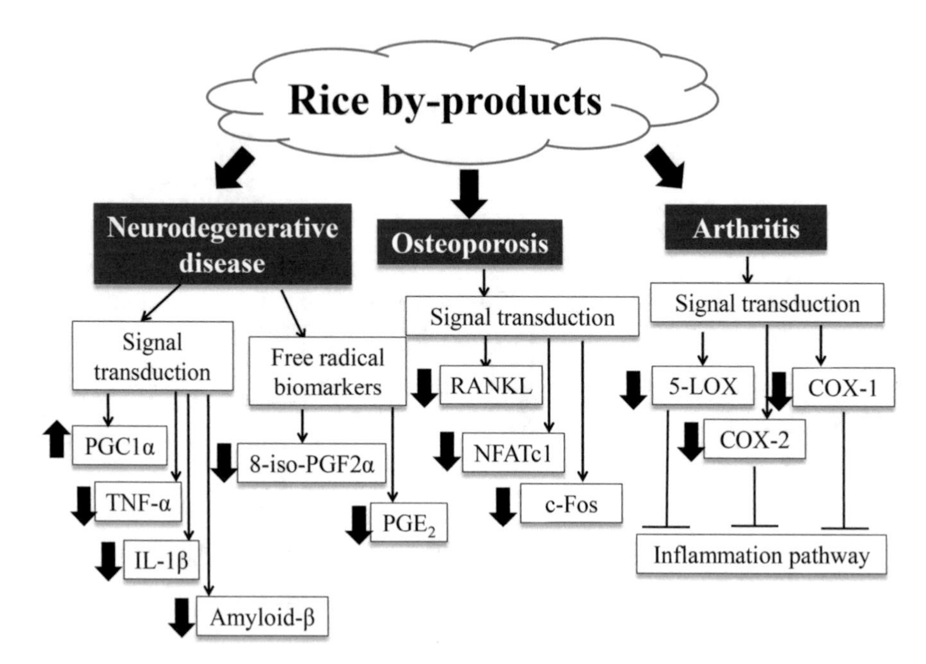

Fig. 7.3 The mechanisms of rice by-products in neurodegenerative disease, osteoporosis, and arthritis. *COX* cyclooxygenase, *IL-1β* interleukin-1beta, *NFATc1* nuclear factor of activated T-cells, cytoplasmic 1, *PGC1α* peroxisome proliferator-activated receptor gamma coactivator 1-alpha, *PGE_2* prostaglandin E2, *RANKL* receptor activator of nuclear factor κB ligand, *TNF-α* tumor necrosis factor-alpha, *5-LOX* 5-lipoxygenase

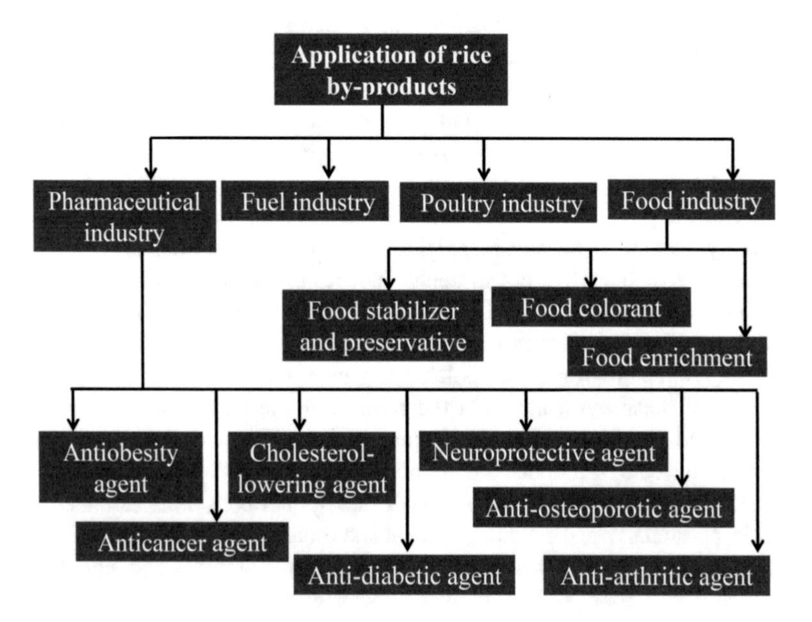

Fig. 7.4 Application of rice by-products in various industries

utilization of rice bran for edible oil extraction will potentially add value to the rice milling sector. Interestingly, rice by-products not only contributed to the development of functional foods and value-added foods that are currently high in demands but also utilized in the fuel industry and alternative poultry production (Fig. 7.4), suggesting the numerous potentials of rice by-products. Therefore, application of these rice by-products in various industries is essential to improve efficiency and reduce waste (Bharathiraja et al. 2017). Taken together, rice by-products seem a good dietary agent for providing nutrients and treatment of metabolic ailments. However, the potential of bioactive compounds in certain diseases requires further elucidation.

References

Aalim H, Belwal T, Wang Y et al (2019) Purification and identification of rice bran (Oryza sativa L.) phenolic compounds with *in-vitro* antioxidant and antidiabetic activity using macroporous resins. Int J Food Sci Technol 54:715–722

Al-Okbi S, Mohamed DA, Hamed TE et al (2020) Rice bran as source of nutraceuticals for management of cardiovascular diseases, cardio-renal syndrome and hepatic cancer. J Herbmed Pharmacol 9:68–74

Bharathiraja S, Suriya J, Krishnan M et al (2017) Chapter six—production of enzymes from agricultural wastes and their potential industrial applications. Adv Food Nutr Res 80:125–148

Deng G-F, Xu X-R, Zhang Y et al (2013) Phenolic compounds and bioactivities of pigmented rice. Crit Rev Food Sci Nutr 53:296–306

Dokkaew A, Punvittayagul C, Insuan O et al (2019) Protective effects of defatted sticky rice bran extracts on the early stages of hepatocarcinogenesis in rats. Molecules 24:2142

Esa NM, Ling TB, Peng LS (2013) By-products of rice processing: an overview of health benefits and applications. J Rice Res 1:107

Felhi S, Daoud A, Hajlaoui H et al (2017) Solvent extraction effects on phytochemical constituents profiles, antioxidant and antimicrobial activities and functional group analysis of *Ecballium elaterium* seeds and peels fruits. Food Sci Technol 37:483–492

Gul K, Singh AK, Jabeen R (2016) Nutraceuticals and functional foods: the foods for the future world. Crit Rev Food Sci Nutr 56:2617–2627

Meselhy KM, Shams MM, Sherif NH et al (2019) Phenolic profile and *in vivo* cytotoxic activity of rice straw extract. Pharm J 11:849–857

Panche AN, Diwan AD, Chandra SR (2016) Flavonoids: an overview. J Nutr Sci 5:e47

Peanparkdee M, Iwamoto S (2019) Bioactive compounds from by-products of rice cultivation and rice processing: extraction and application in the food and pharmaceutical industries. Trends Food Sci Technol 86:109–117

Ryan EP (2011) Bioactive food components and health properties of rice bran. J Am Vet Med Assoc 238:593–600

Tan BL, Norhaizan ME (2017) Scientific evidence of rice by-products for cancer prevention: chemopreventive properties of waste products from rice milling on carcinogenesis *in vitro* and *in vivo*. Biomed Res Int 2017:9017902. 18p

Tan BL, Norhaizan ME, Rahman HS et al (2014) Brewers' rice induces apoptosis in azoxymethane-induced colon carcinogenesis in rats via suppression of cell proliferation and the Wnt signaling pathway. BMC Complement Altern Med 14:304

Tan BL, Norhaizan ME, Huynh K et al (2015) Water extract of brewers' rice induces apoptosis in human colorectal cancer cells via activation of caspase-3 and caspase-8 and downregulates the Wnt/β-catenin downstream signaling pathway in brewers' rice-treated rats with azoxymethane-induced colon carcinogenesis. BMC Complement Altern Med 15:205

Tan BL, Norhaizan ME, Liew W-P-P (2018) Nutrients and oxidative stress: Friend or foe? Oxid Med Cell Longev 2018:9719584. 24p

Triratanasirichai K, Singh M, Anal AK (2017) Value-added byproducts from rice processing industries. In: Anal AK (ed) Food processing by-products and their utilization. Wiley and Sons, Hoboken, pp 277–293

Conclusion

This book has provided substantial evidence that consumption of rice by-products and its derived compounds may provide the optimal health both *in vivo* and *in vitro* models as well as human studies. In this regard, rice by-products and its derived components hold great promising and may play a crucial role in ameliorating chronic diseases, inhibiting inflammation, improving immune system, as well as scavenging ROS. The broad spectrum of processes in which the phytochemicals are involved suggests that a protective role of rice by-products in the pathogenesis of chronic diseases. The global affordability and availability of rice by-products may provide a better public health opportunity in both developing and developed countries. Overall, this book may pave the way for the potential use of brewers' rice, rice germ, rice husk, rice bran, and broken rice as a functional ingredient in various foods. The potential implication of rice by-products in relation to different diseases could be significant and is warranted to be evaluated in randomized clinical trials.

Competing Interests

The authors declare that there is no conflict of interest regarding the publication of this paper.

B. L. Tan, M. E. Norhaizan, *Rice By-products: Phytochemicals and Food Products Application*, https://doi.org/10.1007/978-3-030-46153-9

Index

B. L. Tan, M. E. Norhaizan, *Rice By-products: Phytochemicals and Food Products Application*, https://doi.org/10.1007/978-3-030-46153-9

Printed in the United States
by Baker & Taylor Publisher Services